Kähler Spaces, Nilpotent Orbits, and Singular Reduction

MEMOIRS
of the
American Mathematical Society

Number 814

Kähler Spaces, Nilpotent Orbits, and Singular Reduction

Johannes Huebschmann

November 2004 • Volume 172 • Number 814 (third of 4 numbers) • ISSN 0065-9266

American Mathematical Society
Providence, Rhode Island

2000 *Mathematics Subject Classification.*
Primary 53D20; Secondary 14L24, 14L30, 17B63, 17B65, 17B66, 17B81, 17C36, 17C37, 17C40, 17C70, 17C90, 32C20, 32M15, 32Q15, 32S05, 32S60, 53C30, 53D17, 53D30, 53D50, 81S10.

Library of Congress Cataloging-in-Publication Data

Huebschmann, Johannes, 1950–
Kähler spaces, nilpotent orbits, and singular reduction / Johannes Huebschmann.
p. cm. — (Memoirs of the American Mathematical Society, ISSN 0065-9266 ; no. 814)
"Volume 172, number 814 (third of 4 numbers)."
Includes bibliographical references.
ISBN 0-8218-3572-6 (alk. paper)
1. Poisson manifolds, 2. Poisson algebras. 3. Symplectic geometry. 4. Linear algebraic groups. I. Title. II. Series.
QA3.A57 no. 814
[QA614.3]
510 s–dc22
[516.3′62] 2004054529

Memoirs of the American Mathematical Society

This journal is devoted entirely to research in pure and applied mathematics.

Subscription information. The 2004 subscription begins with volume 167 and consists of six mailings, each containing one or more numbers. Subscription prices for 2004 are $583 list, $466 institutional member. A late charge of 10% of the subscription price will be imposed on orders received from nonmembers after January 1 of the subscription year. Subscribers outside the United States and India must pay a postage surcharge of $31; subscribers in India must pay a postage surcharge of $43. Expedited delivery to destinations in North America $35; elsewhere $130. Each number may be ordered separately; *please specify number* when ordering an individual number. For prices and titles of recently released numbers, see the New Publications sections of the *Notices of the American Mathematical Society.*

Back number information. For back issues see the *AMS Catalog of Publications.*

Subscriptions and orders should be addressed to the American Mathematical Society, P. O. Box 845904, Boston, MA 02284-5904, USA. *All orders must be accompanied by payment.* Other correspondence should be addressed to 201 Charles Street, Providence, RI 02904-2294, USA.

Memoirs of the American Mathematical Society is published bimonthly (each volume consisting usually of more than one number) by the American Mathematical Society at 201 Charles Street, Providence, RI 02904-2294, USA. Periodicals postage paid at Providence, RI. Postmaster: Send address changes to Memoirs, American Mathematical Society, 201 Charles Street, Providence, RI 02904-2294, USA.

This publication is indexed in *Science Citation Index*®, *SciSearch*®, *Research Alert*®, *CompuMath Citation Index*®, *Current Contents*®/*Physical, Chemical & Earth Sciences.*
Printed in the United States of America.

♾ The paper used in this book is acid-free and falls within the guidelines established to ensure permanence and durability.
Visit the AMS home page at http://www.ams.org/

10 9 8 7 6 5 4 3 2 1 09 08 07 06 05 04

Table of contents

Abstract

For a stratified symplectic space, a suitable concept of stratified Kähler polarization encapsulates Kähler polarizations on the strata and the behaviour of the polarizations across the strata and leads to the notion of stratified Kähler space which establishes an intimate relationship between nilpotent orbits, singular reduction, invariant theory, reductive dual pairs, Jordan triple systems, symmetric domains, and pre-homogeneous spaces: The closure of a holomorphic nilpotent orbit or, equivalently, the closure of the stratum of the associated pre-homogeneous space of parabolic type carries a (positive) normal Kähler structure. In the world of singular Poisson geometry, the closures of principal holomorphic nilpotent orbits, positive definite hermitian JTS's, and certain pre-homogeneous spaces appear as different incarnations of the same structure. The closure of the principal holomorphic nilpotent orbit arises from a semisimple holomorphic orbit by contraction. Symplectic reduction carries a positive Kähler manifold to a positive normal Kähler space in such a way that the sheaf of germs of polarized functions coincides with the ordinary sheaf of germs of holomorphic functions. Symplectic reduction establishes a close relationship between singular reduced spaces and nilpotent orbits of the dual groups. Projectivization of holomorphic nilpotent orbits yields exotic (positive) stratified Kähler structures on complex projective spaces and on certain complex projective varieties including complex projective quadrics. The space of (in general twisted) representations of the fundamental group of a closed surface in a compact Lie group or, equivalently, a moduli space of central Yang-Mills connections on a principal bundle over a surface, inherits a (positive) normal (stratified) Kähler structure. Physical examples are provided by certain reduced spaces arising from angular momentum zero.

Received by the editor September 5, 2002.

2000 *Mathematics Subject Classification.* Primary 53D20; Secondary 14L24 14L30 17B63 17B65 17B66 17B81 17C36 17C37 17C40 17C70 17C90 32C20 32M15 32Q15 32S05 32S60 53C30 53D17 53D30 53D50 81S10.

Key words and phrases. Poisson manifold, Poisson algebra, symplectic reduction, Kähler reduction, stratified Kähler space, holomorphic nilpotent orbit, normal complex analytic space, exotic projective variety, constrained system, J(ordan) T(riple) S(ystem), reductive dual pair, invariant theory, real Lie algebra of hermitian type, pre-homogeneous space, contraction of semisimple holomorphic orbit, moduli space of central Yang-Mills connections, moduli space of semistable holomorphic vector bundles, angular momentum, harmonic oscillator.

Introduction

Given a quantizable system with constraints, the issue arises whether reduction after quantization coincides with quantization after reduction. Reduction after quantization makes sense provided the unreduced phase space is a quantizable smooth symplectic manifold in such a way that the symmetries can be quantized as well; it may then be studied within the usual framework of geometric quantization. Up to now, the available methods have been insufficient to attack the problem of quantization of reduced observables, though, once the reduced phase space is no longer a smooth manifold; we will refer to this situation as the *singular case*. The singular case is the rule rather than the exception. For example, simple classical mechanical systems and the solution spaces of classical field theories involve singularities; see e. g. [**1**] and the references there. In the presence of singularities, the naive restriction of the quantization problem to a smooth open dense part, the "top stratum", may lead to a loss of information and in fact to inconsistent results; see what is said below. Trying to overcome these difficulties on the classical level, we were led to isolate a certain class of "Kähler spaces with singularities", to which the present paper is devoted. Examples thereof arise from symplectic reduction, applied to Kähler manifolds. Other examples arise from certain moduli spaces, of central Yang-Mills connections over a closed surface as well as (equivalently) of suitable representations of the fundamental group of a closed surface in a compact Lie group. Still other examples arise from taking the closure of a holomorphic nilpotent orbit in a real semisimple (or reductive) Lie algebra of hermitian type (see what is said below).

A typical example of the kind of Kähler space with singularities we will study is the complex plane with its ordinary complex analytic structure, but with a Poisson algebra of continuous functions which has the origin as a singularity so that geometrically this plane should then be viewed as a half-cone rather than the ordinary plane; the cone point, i. e. the singularity, cannot be detected by the complex analytic structure. Albeit looking like a toy example, it illustrates the general phenomenon that, under such circumstances, playing off against each other the complex and real (semi-analytic) structures yields geometric insight. Actually, this "exotic" plane is the closure in $\mathfrak{sp}(1,\mathbb{R})$ of what we call a *holomorphic* nilpotent orbit: A real semisimple Lie algebra $\mathfrak{g}$ with Cartan decomposition $\mathfrak{g}=\mathfrak{k}\oplus\mathfrak{p}$ together with an element z in the center of $\mathfrak{k}$, referred to as an *H-element*, such that the restriction of $\mathrm{ad}(z)$ to $\mathfrak{p}$ yields a complex structure on the latter is said to be of *hermitian type* [**79**]; the classification of semisimple Lie algebras of hermitian type plainly parallels E. Cartan's classification of semisimple hermitian symmetric spaces. We will say that an orbit $\mathcal{O}$ in such a Lie algebra $\mathfrak{g}$ is *holomorphic* provided the projection from $\mathcal{O}$ to $\mathfrak{p}$ is a diffeomorphism onto its image in such a way that the resulting complex structure $J_{\mathcal{O}}$ on $\mathcal{O}$ combines with its (Kostant-Kirillov-Souriau) symplectic structure to a positive Kähler structure. Here $\mathfrak{g}$ is identified with its dual by means of an appropriate positive multiple of the Killing form, made precise later in the paper. The holomorphicity of an orbit does *not* depend on the choice of H-element or, equivalently, on the Cartan decomposition; however, the induced complex structure on the orbit *does* depend on that choice; see (3.7.5) and (3.7.6) below. The name "holomorphic" is intended to hint at the fact that the holomorphic discrete series representations arise from holomorphic quantization on integral semisimple holomorphic orbits $\mathcal{O}$ (the requisite complex structure being different from $J_{\mathcal{O}}$, though, see what is said below). Indeed, for any real positive λ, the G-orbit in $\mathfrak{g}$ generated by λz is semisimple and holomorphic; it realizes the smooth manifold underlying the corresponding *symmetric domain* as a symplectic manifold. In Section 3 we will prove that a simple Lie algebra of hermitian type and split rank r has exactly $r+1$ *holomorphic nilpotent* orbits $\mathcal{O}_0, \dots, \mathcal{O}_r$ which are linearly ordered in such a way that $\{0\} = \mathcal{O}_0 \subseteq \overline{\mathcal{O}_1} \subseteq \ldots \subseteq \overline{\mathcal{O}_r}$ and that the projection from $\overline{\mathcal{O}_r}$ to the constituent $\mathfrak{p}$ in the Cartan decomposition is a homeomorphism. This entails the fact that every orbit in the closure $\overline{\mathcal{O}}$ of a holomorphic nilpotent orbit is itself holomorphic and that the projection from $\overline{\mathcal{O}}$ to $\mathfrak{p}$ is a homeomorphism onto its image which, via the complex structure on $\mathfrak{p}$ induced by the H-element z (which is part of the hermitian type structure), turns $\overline{\mathcal{O}}$ into a complex affine variety. The "exotic" plane mentioned earlier is a very special case thereof. In general, $\overline{\mathcal{O}_r}$ may be seen as arising from the corresponding semisimple holomorphic orbits by contraction; this contraction does not change the complex analytic structure. A comment about the complex structures is perhaps in order: The complex structure $J_{\mathcal{O}}$ on a semisimple holomorphic orbit $\mathcal{O}$ arises from the projection to $\mathfrak{p}$ as does the complex structure on $\overline{\mathcal{O}_r}$, whence complex analytically these spaces are just a copy of $\mathfrak{p}$. However, for a semisimple orbit, this is *not* the hermitian symmetric space complex structure.

For intelligibility we will briefly explain a special case, which will serve as the *guiding example*. Let $\ell \geq 1$, and consider the real Lie algebra $\mathfrak{sp}(\ell, \mathbb{R})$. Within $\mathfrak{sl}(2\ell, \mathbb{R})$, we may realize $\mathfrak{sp}(\ell, \mathbb{R})$ as the Lie subalgebra of matrices $\begin{bmatrix} A & B \\ C & -A^{\mathrm{t}} \end{bmatrix}$ where B and C are symmetric $(\ell \times \ell)$-matrices, where A is an arbitrary $(\ell \times \ell)$-matrix, and where A^{t} is the transpose of A. Let $J = \begin{bmatrix} 0 & I \\ -I & 0 \end{bmatrix} \in \mathfrak{sp}(\ell, \mathbb{R})$; then $\frac{1}{2} J$ is an H-element. The corresponding Cartan decomposition has the form

$$\mathfrak{sp}(\ell, \mathbb{R}) = \mathfrak{u}(\ell) \oplus \mathrm{S}^2_{\mathbb{C}}[\mathbb{C}^\ell]$$

where the copy of the unitary Lie algebra $\mathfrak{u}(\ell)$ consists of the matrices $\begin{bmatrix} A & B \\ -B & A \end{bmatrix}$ with A skew symmetric and B symmetric and where the copy $\mathrm{S}^2_{\mathbb{C}}[\mathbb{C}^\ell]$ of the space of complex symmetric $(\ell \times \ell)$-matrices consists of the matrices $\begin{bmatrix} A & B \\ B & -A \end{bmatrix}$ with both A and B real symmetric $(\ell \times \ell)$-matrices. For $0 \leq t \leq \ell$, let

$$I_{\ell,t} = \mathrm{diag}(1, \dots, 1, 0, \dots, 0),$$

the diagonal matrix having t 1's and $\ell - t$ zero's along the diagonal. The (nilpotent) matrix $\begin{bmatrix} 0 & I_{\ell,t} \\ 0 & 0 \end{bmatrix}$ generates a holomorphic nilpotent orbit, indeed, when the above notation is maintained, this is precisely the orbit $\mathcal{O}_t$ and, as t varies between zero and ℓ, this yields all holomorphic nilpotent orbits in $\mathfrak{sp}(\ell, \mathbb{R})$. Under the projection to the symmetric constituent $\mathfrak{p} = \mathrm{S}^2_{\mathbb{C}}[\mathbb{C}^\ell]$ of the Cartan decomposition, the orbit $\mathcal{O}_t$ is mapped diffeomorphically onto the space of complex symmetric $(\ell \times \ell)$-matrices of rank t. Furthermore, the projection is a homeomorphism from the (topological) closure $\overline{\mathcal{O}_\ell}$ of $\mathcal{O}_\ell$ onto $\mathrm{S}^2_{\mathbb{C}}[\mathbb{C}^\ell]$ and, in this fashion, for $0 \leq t \leq \ell$, the closure $\overline{\mathcal{O}_t}$ of $\mathcal{O}_t$ is identified with the space of complex symmetric $(\ell \times \ell)$-matrices of rank at most t. For $\ell = 1$, $\overline{\mathcal{O}_1}$ is the "exotic" plane mentioned before. More details may be found in (3.5.1) below.

Our Kähler spaces with singularities are stratified symplectic spaces with additional structure. Reduced spaces stratified into smooth symplectic manifolds (which arise as symmetry types) occur already in [**2**] (where smooth infinite dimensional pieces are not excluded). The notion of *stratified symplectic structure* on a space X which we will use is that of [**85**]; it is a stratification into smooth finite dimensional manifolds (in a sense stronger than that of [**2**]) together with a Poisson algebra $(C^\infty(X), \{\cdot,\cdot\})$ of continuous functions on X which, on each stratum, restricts to a symplectic Poisson algebra of smooth functions, not necessarily consisting of all smooth functions on the stratum. We deliberately write $C^\infty(X)$ even though this algebra need not be an algebra of ordinary smooth functions; the algebra $C^\infty(X)$ is then referred to as a *smooth* structure on X. A *stratified symplectic space* is a stratified space X together with a stratified symplectic structure $(C^\infty(X), \{\cdot,\cdot\})$, referred to henceforth as the *stratified symplectic Poisson algebra* (of the stratified symplectic space). In [**85**] it is shown that the reduced space for a hamiltonian action of a compact Lie group on a smooth symplectic manifold inherits a stratified symplectic structure, the requisite stratified symplectic Poisson algebra being that given in [**1**]; see also [**11**] for related issues. Stratified symplectic structures have been constructed on various moduli spaces [**20, 34–37**] including moduli spaces of central Yang-Mills connections for bundles on a surface and spaces of (in general twisted) representations of a surface group in a compact Lie group (where "twisted" means suitable representations of the universal central extension), and the stratified symplectic structure on such a space yields a description of its symplectic singularity behaviour; see [**39–41**] for overviews. In the present paper we will explore stratified symplectic spaces arising from holomorphic nilpotent orbits and from reduction of Kähler manifolds; in physics, examples come from standard hamiltonian systems in mechanics, in particular from angular momentum (see below). The relationship between angular momentum zero and certain nilpotent orbits in the real symplectic Lie algebras has been observed already in [**60**]. In fact, the nilpotent orbits described explicitly in that paper are precisely the holomorphic ones but this was not observed there.

The Poisson structure of a stratified symplectic space encapsulates the mutual positions of the symplectic structures on the strata. The *new key notion* which we will introduce in this paper is that of *stratified polarization*, which likewise *encapsulates the mutual positions of polarizations on the strata*. We will show that, for holomorphic nilpotent orbits $\mathcal{O}$ of a semisimple Lie algebra of hermitian type, as well as for what we will call antiholomorphic nilpotent orbits, the closure $\overline{\mathcal{O}}$ carries what we will call a *complex analytic stratified Kähler polarization*. Furthermore, for a (positive) Kähler manifold with a hamiltonian action of a compact Lie group G which preserves the Kähler polarization and extends to an action of the complexification of G, the reduced space inherits a complex analytic stratified Kähler polarization. In general, we refer to a space with a complex analytic stratified Kähler polarization as a *complex analytic stratified Kähler space*. We also introduce the somewhat more general notion of *stratified Kähler polarization* (not necessarily complex analytic); on each stratum, this kind of polarization boils down to an ordinary Kähler polarization, and a space equipped with a stratified Kähler polarization will be referred to as a *stratified Kähler space*. While a Kähler polarization on an ordinary smooth symplectic manifold is well known to be equivalent to a complex structure for which the symplectic structure is a Kähler form, we do not know whether a stratified Kähler polarization (in our sense) on a stratified symplectic space necessarily determines a corresponding complex analytic structure.

The stratification of a complex analytic stratified Kähler space is in general finer than the ordinary complex analytic stratification and the stratified symplectic structure then exhibits singularities which are complex analytically spurious; an illustration is the complex plane with a cone point mentioned above. More striking examples are certain *exotic* complex projective spaces. We define an *exotic* complex projective space to be a complex projective space $\mathbb{C}\mathrm{P}^\ell$ with its ordinary complex structure, but with a *non-standard* stratified symplectic structure $(C^\infty(\mathbb{C}\mathrm{P}^\ell), \{\cdot,\cdot\})$ which is compatible with the complex structure where "non-standard" means that the stratified symplectic structure involves at least two strata. To avoid misunderstanding we note that the definition of such an exotic (stratified) Kähler structure is independent of the notion of holomorphic nilpotent orbit. As pointed out by the referee, an interesting question is whether such exotic structures can be classified but we do not know whether and how this may be achieved. In the present paper we will explore systematic classes of examples. In particular, complex projective 3-space has an exotic structure whose singular locus is a Kummer surface, and the stratified Poisson structure detects the Kummer surface *and* its 16 singular points; see 4.3 below. Complex projective 3-space has yet another exotic structure, having as its singular locus a complex quadric; see Section 10 below. Such exotic spaces even arise from classical constrained systems in mechanics: The reduced space Q_ℓ (say) of ℓ harmonic oscillators in some $\mathbb{R}^\ell$ with total angular momentum zero and constant energy is an exotic complex projective space of complex dimension $d = \frac{\ell(\ell+1)}{2} - 1$ in such a way that, for $1 \leq s \leq \ell$, the reduced spaces Q_s (say) of ℓ harmonic oscillators in some $\mathbb{R}^s$ with total angular momentum zero and (same) constant energy embed naturally into $Q_\ell = \mathbb{C}\mathrm{P}^d$ and constitute an ascending chain $Q_1 \subseteq Q_2 \subseteq \cdots \subseteq Q_\ell$ of (compact) complex analytic stratified Kähler spaces which, as complex analytic spaces, are projective determinantal varieties, and such that the $Q_j \setminus Q_{j-1}$ are the strata of the decomposition of any Q_s $(j \leq s)$. These examples are particular cases of systematic classes of exotic stratified Kähler structures on complex projective spaces alluded to already which will be given in the paper. The question whether there exist exotic structures on complex projective spaces is also prompted by the following observation: For a compact Kähler manifold N which is complex analytically a projective variety (i. e. Hodge manifold), with reference to the standard structure on complex projective space (i. e. the Fubini-Study metric), the Kodaira embedding will not in general be symplectic, but complex projective space might still carry an exotic structure which, via the Kodaira embedding, restricts to the Kähler structure on N. More generally, given a Hodge manifold N together with an appropriate group of symmetries and momentum mapping, reduction carries it to a complex analytic stratified Kähler space N^{red} which is as well a projective variety, and the question arises whether in such a situation the requisite complex projective space carries an exotic structure which, via the Kodaira embedding, restricts to the complex analytic stratified Kähler structure on N^{red}. Our approach in particular demonstrates that this situation actually occurs in "mathematical nature". The particular case $Q_s \subseteq \mathbb{C}\mathrm{P}^d$ $(1 \leq s \leq \ell,\ d = \frac{\ell(\ell+1)}{2} - 1)$ arising from angular momentum is a good illustration.

We close the introduction with a guide through the paper. In Section 1 we reproduce some material from the relationship between Poisson algebras and Lie-Rinehart algebras. In Section 2 we introduce stratified polarized spaces. The aim of Section 3 is to show that *the closure of any holomorphic nilpotent orbit carries a normal complex analytic stratified Kähler structure*; see Theorem 3.2.1 for details. We will, furthermore, give a *complete classification of all holomorphic nilpotent orbits, and we will explicitly describe the real (semi-algebraic) and the complex analytic structure of the closure of any such holomorphic nilpotent orbit*. We will also classify what we call *pseudoholomorphic* nilpotent orbits; a pseudoholomorphic nilpotent orbit $\mathcal{O}$ is one which under the projection to the constituent $\mathfrak{p}$ of the Cartan decomposition is still mapped diffeomorphically onto its image (but the resulting Kähler structure is not constrained to be positive). Along the way, we obtain several results on holomorphic nilpotent orbits that look interesting in their own right: For a classical Lie algebra $\mathfrak{g}$, let E be its standard representation, and consider the ordinary *associative* algebra $\mathrm{End}(E)$ of endomorphisms thereof. We will show that, in the standard cases ($\mathfrak{g} = \mathfrak{sp}(\ell, \mathbb{R}), \mathfrak{su}(p,q), \mathfrak{so}^*(2n)$), an element X of $\mathfrak{g}$ generates a pseudoholomorphic nilpotent orbit if and only if X^2 is zero in $\mathrm{End}(E)$ and that, likewise, an element X of $\mathfrak{so}(2,q)$ generates a pseudoholomorphic nilpotent orbit if and only if X^3 is zero in $\mathrm{End}(E)$. The Lie algebra $\mathfrak{g}$ sits of course inside $\mathrm{End}(E)$ and an element $X \in \mathfrak{g}$ satisfying $X^2 = 0$ or $X^3 = 0$ is plainly nilpotent (i. e. $\mathrm{ad}(X)$ is a nilpotent endomorphism of $\mathfrak{g}$) but the requirement that $X \in \mathfrak{g}$ satisfy $X^2 = 0$ or $X^3 = 0$ has no direct meaning within $\mathfrak{g}$ itself. Another interesting side result is the observation that, when $\mathfrak{g}$ is simple, *a pseudoholomorphic nilpotent orbit $\mathcal{O}$ is actually holomorphic (or antiholomorphic, which means that $-\mathcal{O}$ is holomorphic) if and only if the projection from the closure $\overline{\mathcal{O}}$ to $\mathfrak{p}$ is a homeomorphism onto its image in $\mathfrak{p}$*; see Theorem 3.2.1. Our results are conclusive, except that we do not give explicit defining equations for the complex analytic structures of the closures of the rank 1 holomorphic nilpotent orbits in the two exceptional cases. In [**89**], given a Lie algebra of hermitian type $(\mathfrak{g}, z)$, a nilpotent orbit $\mathcal{O}$ in $\mathfrak{g}$ is defined to be holomorphic provided it corresponds, under the Kostant-Sekiguchi correspondence [**83**], to an orbit in the holomorphic constituent $\mathfrak{p}^+$ of the complexification $\mathfrak{p}^{\mathbb{C}} = \mathfrak{p}^+ \oplus \mathfrak{p}^-$, and it is then proved, via the Kronheimer flow [**57**], that the restriction of the projection to $\mathfrak{p}$ of such an orbit is a diffeomorphism onto its image, that is to say, that such an orbit is (pseudo)holomorphic in our sense. The approach in the present paper is somewhat different, though: We *define* the property of holomorphicity of an orbit in terms of the projection to $\mathfrak{p}$—this definition is more elementary than that involving the Kostant-Sekiguchi correspondence—and we classify *all* holomorphic nilpotent orbits (in our sense) by means of a geometric description which bypasses the Kostant-Sekiguchi correspondence; in particular, our approach does *not* involve the Kronheimer flow at all and is therefore likely to allow for generalization over other (formally real) fields. It also enables us to show that *the projection from a holomorphic nilpotent orbit $\mathcal{O}$ to $\mathfrak{p}$ extends to a homeomorphism from the closure $\overline{\mathcal{O}}$ onto its image in $\mathfrak{p}$*. We do not know whether this property of holomorphic nilpotent orbits can formally be deduced from properties of the Kronheimer flow since passing to the closure amounts to changing the topological type of the corresponding bundle on S^4. This property of holomorphic nilpotent orbits plainly implies the fact that, for a holomorphic nilpotent orbit, the Kostant-Sekiguchi correspondence is compatible with passing to the closure—this is actually true for any nilpotent orbit in a real semisimple Lie algebra, cf. [**4**]—but is considerably stronger than just the compatibility of the Kostant-Sekiguchi correspondence with respect to the closure. In Section 4 we show that *reduction carries an ordinary (positive) Kähler polarization to a (positive) complex analytic stratified Kähler polarization.* As an application, we prove that the *moduli spaces mentioned earlier inherit normal (complex analytic stratified) Kähler space structures.* Classically, such a structure was obtained only in the non-singular case, where these spaces are smooth Kähler manifolds. Our approach bypasses the usual geometric invariant theory construction (which does not give the requisite stratified symplectic structure) and yields an analytic construction of these moduli spaces. A particular case is the space N of representations of the fundamental group of a Riemann surface of genus two in $\mathrm{SU}(2)$; in view of results of Narasimhan-Ramanan [**70, 71**], this space, realized as the moduli space of semistable holomorphic vector bundles of rank two, degree zero, and trivial determinant, is complex analytically a copy of $\mathbb{C}\mathrm{P}^3$, and the subspace of classes of semistable holomorphic vector bundles which are not stable is a Kummer surface. We will show that, indeed, N carries a normal (complex analytic stratified) Kähler structure having the Kummer surface as its singular locus. This is the exotic structure on $\mathbb{C}\mathrm{P}^3$ mentioned earlier, and the Kummer surface, being viewed as realized in $\mathbb{C}\mathrm{P}^3$ by a Kodaira embedding, inherits its (singular) normal Kähler structure from the embedding. In Section 5 we describe the holomorphic nilpotent orbits (for the classical cases) as reduced spaces for suitable momentum mappings; this involves versions of the *First and Second Main Theorem of Invariant Theory* [**93**]; the Second Main Theorem of Invariant Theory will actually be a consequence of our approach to invariant theory. Our tools are the original observation of Kempf-Ness [**49**] and an extension of the *basic construction* in [**55**]; in that reference, this construction is given only for the case of the classical groups over the complex numbers. We extend this construction to the appropriate real forms of the classical groups. The Kempf-Ness observation is then exploited to identify a holomorphic nilpotent orbit as a symplectic quotient with the corresponding complex categorical quotient (in the sense of geometric invariant theory).

At the risk of breaking the flow of the exposition, we now explain briefly how, under the circumstances of the guiding example, the holomorphic nilpotent orbits and their closures arise from reduction: Consider a system of ℓ particles in $\mathbb{R}^s$ with $\mathrm{O}(s,\mathbb{R})$-symmetry. Thus the unreduced phase space is a product $(\mathrm{T}^*\mathbb{R}^s)^\ell$ of ℓ copies of the total space $\mathrm{T}^*\mathbb{R}^s$ of the cotangent bundle on $\mathbb{R}^s$, endowed with the obvious symplectic structure, and the requisite momentum mapping $\mu\colon (\mathrm{T}^*\mathbb{R}^s)^\ell \to \mathfrak{so}(s,\mathbb{R})$ (the Lie algebra $\mathfrak{so}(s,\mathbb{R})$ being identified with its dual in the standard fashion) is the ordinary angular momentum mapping (the sum of the individual angular momenta). This system has an obvious $\mathrm{Sp}(\ell,\mathbb{R})$-symmetry as well, and the $\mathrm{Sp}(\ell,\mathbb{R})$-momentum mapping

$$\mu_{\mathrm{Sp}(\ell,\mathbb{R})}\colon (\mathrm{T}^*\mathbb{R}^s)^\ell \longrightarrow \mathfrak{sp}(\ell,\mathbb{R})$$

(the Lie algebra $\mathfrak{sp}(\ell,\mathbb{R})$ being likewise identified with its dual in the standard fashion) induces a homeomorphism from the $\mathrm{O}(s,\mathbb{R})$-reduced space onto the closure $\overline{\mathcal{O}_t}$ of the holomorphic nilpotent orbit $\mathcal{O}_t$ where $t = \min\{s,\ell\}$, that is, onto $\overline{\mathcal{O}_s}$ when $s \leq \ell$ and onto $\overline{\mathcal{O}_\ell}$ otherwise. This identification between the $\mathrm{O}(s,\mathbb{R})$-reduced space and the closure of a holomorphic nilpotent orbit is actually an isomorphism of stratified symplectic spaces. Furthermore, the space $(\mathrm{T}^*\mathbb{R}^s)^\ell$ inherits an obvious complex structure which, with the symplectic structure, combines to a Kähler structure, the $\mathrm{O}(s,\mathbb{R})$-symmetry preserves the Kähler structure, and under the Kempf-Ness homeomorphism from the $\mathrm{O}(s,\mathbb{R})$-reduced space onto the categorical $\mathrm{O}(s,\mathbb{C})$-quotient, the reduced space inherits a complex analytic structure which combines with the stratified symplectic structure to a normal Kähler structure. Under the identification with the closure of a holomorphic nilpotent orbit, the normal Kähler structures correspond. See Section 5 below for details; the general construction given there is phrased in terms of certain spaces of homomorphisms, though, not in terms of cotangent bundles. For the special case of angular momentum, the equivalence of the present description in terms of cotangent bundles with the general one is given as an illustration in Section 5 below.

The construction in Section 5 has a consequence for *singular reduction* which is interesting in its own right: Let H be a compact Lie group which may be written as the direct product of copies of classical groups of the kind $\mathrm{O}(s_j,\mathbb{R})$, $\mathrm{Sp}(s_k)$, $\mathrm{U}(s_m)$; for each of these, let E_{s_j}, E_{s_k}, E_{s_m} be the standard unitary representation, $\mathrm{O}(s_j,\mathbb{R})$ being viewed as a subgroup of $U(s_j)$ in the usual way. Pick natural numbers ℓ_j, ℓ_k, ℓ_m, $\overline{\ell_m}$, consider the representations $E_{s_j}^{\ell_j}$, $E_{s_k}^{\ell_k}$, $E_{s_m}^{\ell_m}$, $\overline{E}_{s_m}^{\overline{\ell_m}}$ and, finally, take the sum E of them. This is a unitary H-representation. Here $\overline{E}_{s_m}$ is the conjugate representation of E_{s_m}. The representations E_{s_j} (being real) and E_{s_k} (being quaternionic) are actually self-conjugate whence there is no need to take conjugate representations for the $\mathrm{O}(s_j,\mathbb{R})$'s, nor for the $\mathrm{Sp}(s_k)$'s. Let $\mu\colon E \to \mathfrak{h}$ be the unique H-momentum mapping of E having the value zero at the origin, and let $\mathfrak{g}$ be the real reductive Lie algebra which arises, accordingly, as a sum copies of the $\mathfrak{sp}(\ell_j,\mathbb{R})$'s, $\mathfrak{so}^*(2\ell_k)$'s, $\mathfrak{u}(\ell_m,\overline{\ell_m})$'s. Corollary 5.4.4 below says that, under these circumstances, *the H-reduced space E^{red} is as a normal Kähler space isomorphic to the closure of a holomorphic nilpotent orbit in $\mathfrak{g}$*. Furthermore, *every holomorphic nilpotent orbit in $\mathfrak{g}$ arises in this way*. It is interesting to point out, cf. [**28**], that the groups H and the direct product G of the corresponding copies of $\mathrm{Sp}(\ell_j,\mathbb{R})$, $\mathrm{SO}^*(2\ell_k)$, $\mathrm{U}(\ell_m,\overline{\ell_m})$, constitute a real *reductive dual pair* in $\mathrm{Sp}(E)$, the symplectic structure on E being determined by its unitary one. More generally, let H be a Lie group which may be written as the direct product of copies of classical groups of the kind $\mathrm{O}(s_j',s_j'')$, $\mathrm{Sp}(s_k',s_k'')$, $\mathrm{U}(s_m',s_m'')$, and let E be an H-representation which is a sum of standard representations. In view of [**28**], the groups G and H then still constitute a reductive dual pair in $\mathrm{Sp}(E)$, the symplectic structure on E being the obvious one on E determined by the data (see Section 5 below for details). Let $\mu\colon E \to \mathfrak{h}$ be the unique H-momentum mapping of E having the value zero at the origin. Corollary 5.5.4 below says that *the H-reduced space E^{red}* (which we define to be that of *closed* H-orbits in $\mu^{-1}(0)$) *is as a stratified space isomorphic to a space N which is a union of pseudoholomorphic nilpotent orbits in a Lie algebra $\mathfrak{g}$ of the above kind in such a way that the map from E^{red} to N is a Poisson map. Furthermore, every pseudoholomorphic nilpotent orbit of $\mathfrak{g}$ arises in this way (as a stratum of a reduced space of the kind E^{red}).*

In Section 6 we give a construction which yields the holomorphic nilpotent orbits of $\mathfrak{so}(2,q)$ by singular reduction with respect to the non-compact group $\mathrm{Sp}(1,\mathbb{R})$; for arbitrary q, it does not seem possible to obtain these orbits by singular reduction with respect to a compact group, though. In Section 7 we relate holomorphic nilpotent orbits with positive definite hermitian J(ordan) T(riple) S(ystem)s, and in Section 8 we give some relevant information for the exceptional cases in terms of JTS's. To explain what we achieve at that stage, let $(\mathfrak{g}, z)$ be a real reductive Lie algebra of hermitian type, with Cartan decomposition $\mathfrak{g} = \mathfrak{k} \oplus \mathfrak{p}$ and complexification $\mathfrak{g}_{\mathbb{C}} = \mathfrak{p}^- \oplus \mathfrak{k}_{\mathbb{C}} \oplus \mathfrak{p}^+$, and let G and K denote the adjoint group and maximal compact subgroup with $\mathrm{Lie}(K) = \mathfrak{k}$, respectively. We will establish the following facts, cf. Theorem 7.3 for the classical cases and Theorems 8.4.1 and 8.4.2 for the exceptional ones: *Under the homeomorphism (induced by the projection from $\mathfrak{g}$ to $\mathfrak{p}$) from the closure $\overline{\mathcal{O}}$ of the principal holomorphic nilpotent orbit $\mathcal{O}$ onto the complex vector space $\mathfrak{p}^+$, the G-orbit stratification of $\overline{\mathcal{O}}$ passes to the stratification of $\mathfrak{p}^+$ by Jordan rank, with reference to the JTS-structure on $\mathfrak{p}^+$, and the latter stratification, in turn, coincides with the $K^{\mathbb{C}}$-orbit stratification of $\mathfrak{p}^+$. In this fashion, the induced stratified symplectic Poisson structure on $\mathfrak{p}^+$ turns the latter into a normal complex analytic stratified Kähler space, and the Poisson structure detects the stratification by Jordan rank.* This reduces the study of holomorphic nilpotent orbits in $\mathfrak{g}$ to that of the hermitian Jordan triple system $\mathfrak{p}^+$: *All geometric information of the holomorphic nilpotent orbits is encoded in that* JTS, *even the stratified symplectic Poisson structure since the latter is determined by the Lie bracket in $\mathfrak{g}$, which can be reconstructed from the* JTS; this parallels the well known fact that all geometric information of a symmetric domain is encoded in its JTS, cf. e. g. [**65**]. Our results entail in particular that the pre-homogeneous spaces studied and classified in [**68**]—these are precisely those which arise from positive definite hermitian JTS's—carry more structure; indeed, such a space necessarily underlies the closure of the principal holomorphic nilpotent orbit (viewed as a complex analytic stratified Kähler space) of a semisimple Lie algebra of hermitian type. As a byproduct, we obtain the Bernstein-Sato polynomials for the regular pre-homogeneous spaces under discussion; see Theorems 7.1 and 8.4.1 for details. For example, for $\mathfrak{e}_{7(-25)}$, this polynomial is Freudenthal's generalized (3×3)-determinant over a complex split octonion algebra [**15**] or, equivalently, Jacobson's generic norm. In Section 9 we show how the *closure of the principal holomorphic nilpotent orbit arises from the corresponding semisimple holomorphic orbits by contraction.* In Section 10 we projectivize the holomorphic nilpotent orbits by means of *stratified symplectic reduction* with respect to the circle group. This gives systematic classes of examples of *exotic projective spaces*. In particular, the principal holomorphic nilpotent orbit in $\mathfrak{so}(2, q)$ is a copy of $\mathbb{C}^q$ having as singular locus in the sense of stratified Kähler spaces an affine quadric, and projectivization thereof yields the *exotic projective space* $\mathbb{C}\mathrm{P}^{q-1}$ having as *singular locus a projective quadric*. Thus the projective quadric inherits a Kähler, in fact Hodge structure whose symplectic constituent does *not* arise (via the embedding) from the Fubini-Study metric on projective space, and the underlying real space arises from the non-zero non-principal holomorphic nilpotent orbit in $\mathfrak{so}(2, q)$ by projectivization. This is a particular example of a Kodaira embedding of the kind mentioned earlier, where the symplectic structure results from a proper stratified symplectic Poisson algebra on projective space. For the exceptional Lie algebra $\mathfrak{e}_{7(-25)}$, we obtain an *exotic Kähler structure* on $\mathbb{C}\mathrm{P}^{26}$ having as *singular locus the projective cubic* defined by the generic norm over the octonions mentioned earlier, and this cubic hypersurface, in turn, inherits a complex analytic stratified Kähler structure with two strata whose underlying real structure arises from the closure of the rank 2 holomorphic nilpotent orbit in $\mathfrak{e}_{7(-25)}$ by projectivization. Actually, this cubic has played a major role in the development of the exceptional Lie groups; a version of it was studied by E. Cartan. In the standard cases ($\mathfrak{g} = \mathfrak{sp}(\ell, \mathbb{R}), \mathfrak{su}(p, q), \mathfrak{so}^*(2n)$), via the principle of reduction in stages, the projectivized holomorphic nilpotent orbits may also be obtained by ordinary symplectic reduction, as explained in Theorem 10.1; somewhat amazingly, this does not seem to be possible for $\mathfrak{so}(2, q)$ for general q, nor for $\mathfrak{e}_{6(-14)}$ or $\mathfrak{e}_{7(-25)}$; in these cases, we obtain the projectivized holomorphic nilpotent orbits only by *proper stratified symplectic reduction*, with respect to the circle group. Finally, in Section 11 we compare our notion of stratified Kähler space with that of Kähler space with singularities introduced by GRAUERT [**16**] and with subsequent refinements thereof [**88**], referred to in [**22**] as *stratified Kählerien spaces*. There is a substantial difference between complex analytic stratified Kähler spaces in our sense and stratified Kählerien spaces, though. Our emphasis is on the Poisson structure, and this is crucial for issues related with quantization; stratified Kählerien spaces do not involve Poisson structures at all.

Thus the notion of stratified Kähler space establishes an intimate relationship between nilpotent orbits, singular reduction, invariant theory, reductive dual pairs, Jordan triple systems, symmetric domains, and pre-homogeneous spaces; in particular, in the world of singular Poisson geometry, the closures of principal holomorphic nilpotent orbits, positive definite hermitian JTS's, and certain pre-homogeneous spaces appear as different incarnations of the same structure and, in the classical cases, that structure arises, via invariant theory and singular reduction, from a non-singular one phrased in terms of dual pairs. This Poisson geometry is, perhaps, already lurking behind LIE's theory of *function groups*, cf. the notion of dual pair of Poisson mappings in [**91**]. Somewhat amazingly, Jordan algebras (and generalizations thereof: JTS's)—which have been invented in search for new models of quantum mechanics (cf. e. g. what is said in [**65**])—are now seen to illuminate certain classically reduced phase spaces (e. g. that of finitely many harmonic oscillators with total angular momentum zero). The length of the paper is to some extent explained by the need for various different descriptions of holomorphic nilpotent orbits: To obtain a complete list thereof, the relative root systems suffice; to see that the list is complete, though, we need the classification of nilpotent orbits in the literature in terms of classical matrix realizations, matrix realizations are needed as well to explain the relationship with singular reduction and, at times, structural insight is more easily obtained from matrix realizations than from root systems. To describe the link with JTS's—the exceptional Lie algebras are actually constructed from JTS's—we reproduce a description of the requisite JTS's taylored to our purposes.

We conclude with a somewhat vague comment about normality, a notion which so far does not seem to have played a major role in symplectic or Poisson geometry. The results of the present paper indicate that, perhaps, normal Kähler spaces constitute the *correct* category for doing Kähler reduction. While smoothness is not in general preserved, in the circumstances studied in our paper, normality *is* preserved under reduction. What remains to be done is to develop a general reduction procedure for (in general singular) normal Kähler spaces.

In a follow-up paper [**43**], we have extended the ordinary holomorphic quantization scheme to stratified Kähler spaces. Using this extension, we have established the fact that, indeed, in a suitable sense, *in the framework of Kähler quantization, reduction after quantization is equivalent to quantization after reduction, observables included.* In particular, in that paper, explicit examples are given which justify the claim made earlier that, in the presence of singularities, restricting quantization to a smooth open dense part (the top stratum) leads in general to a *loss of information and in fact to inconsistent results*. In another follow-up paper [**44**], we have shown that the geometry of certain *Severi* varieties may be elucidated in terms of holomorphic nilpotent orbits.

I am much indebted to A. Weinstein and T. Ratiu for discussions, and for their encouragement to carry out the research program the present paper is part of. I am much indebted as well to M. Duflo for having introduced me into SATAKE's book [**79**] and for having suggested a possible relationship between certain nilpotent orbits and Jordan algebras. Thanks are also due to R. Buchweitz, R. Cushman, J. Hilgert, F. Hirzebruch, F. Lescure, O. Loos, L. Manivel, J. Marsden, E. Neher, J. Sekiguchi, T. Springer, J. Stasheff, M. Vergne, R. Weissauer and last but not least to the referee, for support at various stages of the project. I owe a special debt to SATAKE's book [**79**]; without this book, the paper would not have been materialized in its present form.

The paper is dedicated to the memory of Peter Slodowy. At an early stage of the project, he encouraged me to pursue certain questions related with the mutual positions of those nilpotent orbits which are called *holomorphic* in the present paper and, as the project grew, he provided moral support at various stages. Last year, Peter Slodowy died after a long illness.

1. Poisson algebras and Lie-Rinehart algebras

Let R be a commutative ring with 1 taken as ground ring which, for the moment, may be arbitrary. For a commutative R-algebra A, we denote by $\mathrm{Der}(A)$ the R-Lie algebra of derivations of A, with its standard Lie algebra structure. Recall that a *Lie-Rinehart algebra* (A, L) consists of a commutative R-algebra A and an R-Lie algebra L together with an A-module structure on L and an L-module structure $L \to \mathrm{Der}(A)$ on A (that is, L acts on A by derivations), and these are required to satisfy two compatibility conditions modeled on the properties of the pair (A, L), A being the ring of smooth functions $C^\infty(N)$ on a smooth manifold N and L the Lie algebra $\mathrm{Vect}(N)$ of smooth vector fields on N. Explicitly, these compatibility conditions read $a(\alpha(b)) = (a\alpha)b$ and $[\alpha, a\beta] = \alpha(a)\beta + a[\alpha, \beta]$, where $a, b \in A$ and $\alpha, \beta \in L$. Given an arbitrary commutative algebra A over R, another example of a Lie-Rinehart algebra is the pair $(A, \mathrm{Der}(A))$, with the obvious action of $\mathrm{Der}(A)$ on A and obvious A-module structure on $\mathrm{Der}(A)$. There is an obvious notion of morphism of Lie-Rinehart algebras and, with this notion of morphism, Lie-Rinehart algebras constitute a category. Given a Lie-Rinehart algebra (A, L) we shall occasionally refer to L as an *(R, A)-Lie algebra*. More details may be found in RINEHART [**78**] and in our papers [**29**] and [**30**].

Let $(A, \{\cdot,\cdot\})$ be a Poisson algebra, and let D_A be the A-module of formal differentials of A. For $u, v \in A$, the assignment $\pi(du, dv) = \{u, v\}$ yields an A-valued 2-form $\pi = \pi_{\{\cdot,\cdot\}}$ on D_A, the *Poisson 2-form* for $(A, \{\cdot,\cdot\})$. Its adjoint

$$\pi^\sharp \colon D_A \longrightarrow \mathrm{Der}(A) = \mathrm{Hom}_A(D_A, A) \tag{1.1'}$$

is a morphism of A-modules, and the formula

$$[adu, bdv] = a\{u, b\}dv + b\{a, v\}du + abd\{u, v\} \tag{1.2}$$

yields a Lie bracket $[\cdot,\cdot]$ on D_A, viewed as an R-module. The A-module structure on D_A, the bracket $[\cdot,\cdot]$, and the morphism $\pi^\sharp$ of A-modules endow the pair (A, D_A) with a Lie-Rinehart algebra structure in such a way that $\pi^\sharp$ is a morphism of Lie-Rinehart algebras. See [**29**] (3.8) for details. We write $D_{\{\cdot,\cdot\}} = (D_A, [\cdot,\cdot], \pi^\sharp)$ and, for $\alpha \in D_A$, we write $\alpha^\sharp = \pi^\sharp(\alpha) \in \mathrm{Der}(A)$.

We now take as ground ring that of the reals $\mathbb{R}$ or that of the complex numbers $\mathbb{C}$; we shall occasionally use the neutral notation $\mathbb{K}$ for either of them. We shall consider spaces N with an algebra of continuous $\mathbb{K}$-valued functions, deliberately denoted by $C^\infty(N)$, for example ordinary smooth manifolds and ordinary smooth functions; such an algebra $C^\infty(N)$ will then be referred to as a *smooth* structure on N and will be viewed as part of the structure, and the pair $(N, C^\infty(N))$ will be referred to as a *smooth* space. A space may support *distinct* smooth structures, though. Given a space N with a smooth structure $C^\infty(N)$, we shall write $\Omega^1(N)$ for the space of formal differentials with those differentials divided out that are *zero at each point*, cf. [**56**]. More precisely: Let $p \in N$, and view $\mathbb{K}$ as a $C^\infty(N)$-module by means of the evaluation map which assigns to a function f its value $f(p)$ at p; we shall occasionally denote this $C^\infty(N)$-module by $\mathbb{K}_p$. A formal differential is *zero at the point p* of N provided it passes to zero in $(D_{C^\infty(N)}) \otimes_{C^\infty(N)} \mathbb{K}_p$. For example, when N is an ordinary smooth manifold and f an ordinary smooth function, in local coordinates $x_1, \ldots, x_m$, a formal differential of the kind $df - \sum \frac{\partial f}{\partial x_j} dx_j$ is non-zero but is zero at each point. For a general smooth space $(N, C^\infty(N))$, the induced

$C^\infty(N)$-module morphism

$$\mathrm{Hom}_{C^\infty(N)}(\Omega^1(N), C^\infty(N)) \to \mathrm{Hom}_{C^\infty(N)}(D_{C^\infty(N)}, C^\infty(N)) = \mathrm{Der}(C^\infty(N))$$

is clearly an isomorphism. At a point p of N, the object $\Omega^1(N)$ amounts to the ordinary space of differentials, in the following sense: Denote by $\mathfrak{m}_p$ the ideal of functions in $C^\infty(N)$ that vanish at the point p. The $\mathbb{K}$-linear map which assigns to a function f in $\mathfrak{m}_p$ its formal differential induces isomorphisms

$$\mathfrak{m}_p/\mathfrak{m}_p^2 \to (D_{C^\infty(N)}) \otimes_{C^\infty(N)} \mathbb{K}_p \to \Omega^1(N) \otimes_{C^\infty(N)} \mathbb{K}_p$$

of $\mathbb{K}$-vector spaces, and each of these may be taken as the space of *differentials at* p; over the reals, this space, in turn, may be identified with the *cotangent space* T_p^*N. When N is an ordinary smooth manifold, $\Omega^1(N)$ amounts to the space of smooth sections of the cotangent bundle. For a general smooth space N over the reals, when $A = C^\infty(N) = C^\infty(N, \mathbb{R})$ is endowed with a Poisson structure, the formula (1.2) yields a Lie-bracket $[\cdot,\cdot]$ on the $C^\infty(N)$-module $\Omega^1(N)$ and the 2-form $\pi_{\{\cdot,\cdot\}}$ is still defined on $\Omega^1(N)$; its adjoint then yields an $C^\infty(N)$-linear map

$$\pi^\sharp \colon \Omega^1(N) \longrightarrow \mathrm{Hom}_{C^\infty(N)}(\Omega^1(N), C^\infty(N)) \stackrel{\cong}{\longrightarrow} \mathrm{Der}(C^\infty(N)). \tag{1.1}$$

PROPOSITION 1.3. *The $C^\infty(N)$-module structure on $\Omega^1(N)$, the bracket $[\cdot,\cdot]$, and the morphism $\pi^\sharp$ of $C^\infty(N)$-modules endow the pair $(C^\infty(N), \Omega^1(N))$ with a Lie-Rinehart algebra structure in such a way that $\pi^\sharp$ is a morphism of Lie-Rinehart algebras.*

PROOF. Indeed the obvious projection map from $D_{C^\infty(N)}$ to $\Omega^1(N)$ is compatible with the structure. The non-triviality of the kernel of this projection map does not cause any problem. □

We shall write $(\Omega^1(N), [\cdot,\cdot], \pi^\sharp)$ as $\Omega^1(N)_{\{\cdot,\cdot\}}$. When N is an ordinary smooth manifold, the range $\mathrm{Der}(C^\infty(N))$ of the adjoint map $\pi^\sharp$ from $\Omega^1(N)$ to $\mathrm{Der}(C^\infty(N))$ boils down to the space $\mathrm{Vect}(N)$ of smooth vector fields on N. In this case, the Poisson structure on N is symplectic, that is, arises from a (uniquely determined) symplectic structure on N, if and only if $\pi^\sharp$, which may now be written as a morphism of smooth vector bundles from the cotangent bundle to the tangent bundle, is an isomorphism. Thus the 2-form $\pi_{\{\cdot,\cdot\}}$, which is defined for *every* Poisson algebra, generalizes the symplectic form of a symplectic manifold; see Section 3 of **[29]** for details.

2. Stratified polarized spaces

The aim of this section is to develop a satisfactory notion of Kähler polarization for a stratified symplectic space.

Let N be a symplectic manifold, with symplectic Poisson algebra

$$(C^\infty(N), \{\cdot,\cdot\}),$$

and write π for its Poisson 2-form. Since N is symplectic, the complexification

$$\pi^\sharp_{\{\cdot,\cdot\}} \otimes \mathbb{C}\colon \Omega^1(N,\mathbb{C})_{\{\cdot,\cdot\}} \to \mathrm{Vect}(N,\mathbb{C}) \tag{2.1}$$

of the adjoint $\pi^\sharp_{\{\cdot,\cdot\}}$ is an isomorphism of $(\mathbb{C}, C^\infty(N,\mathbb{C}))$-Lie algebras. Hence a complex polarization (i. e. integrable Lagrangian distribution) $F \subseteq \mathrm{T}^{\mathbb{C}}N$ for N corresponds to a certain $(\mathbb{C}, C^\infty(N,\mathbb{C}))$-Lie subalgebra P of $\Omega^1(N,\mathbb{C})_{\{\cdot,\cdot\}}$ which is just the pre-image under (2.1) of the space ΓF of sections of F. The *polarized* complex functions determined by F are smooth complex valued functions f defined on open sets of N satisfying $Zf = 0$ for *every* smooth vector field Z in ΓF; when N is a Kähler manifold and F the holomorphic polarization, these are ordinary holomorphic functions. In general, non-zero polarized functions exist at most locally. Following [**94**], we shall refer to a continuous function f, defined on an open subset of a space X, as a *function in* X. For general N and F, given a function f in N, for a suitable compactly supported smooth bump function u on N (the notion of bump function will be made precise below), the differential udf is defined everywhere on N. Notice that we can extend f to a smooth function $\widetilde f$ on N in such a way that $udf = ud\widetilde f$, so the differential udf really lies in the $C^\infty(N)$-module of differentials, for the algebra $C^\infty(N)$ of smooth functions on the whole space N. Then P is generated by these differentials. A little thought reveals that, for every such f and h, with appropriate bump functions u and v so that the differentials udf and vdh are defined everywhere on N, the Lie bracket (1.2) induces a bracket on P. This bracket is given by the expression

$$[udf, vdh] = u\{f,v\}dh + v\{u,h\}df;$$

notice that the third term $uvd\{f,h\}$ (which has initially to occur, cf. (1.2)) is zero since f and h are polarized.

Let X be a stratified space, and let $C^\infty(X)$ be a smooth structure on X which, on each stratum, restricts to an algebra of ordinary smooth functions; henceforth all smooth spaces coming into play in the paper will be of this kind. Given such a space X and an open subset O thereof, with its induced smooth structure, and a smooth function f on O, we refer to f as a *smooth function in* X. When X is a stratified symplectic space, we will occasionally refer to the stratification as the *symplectic stratification.*

Let X be a stratified symplectic space, with complex Poisson algebra

$$(C^\infty(X,\mathbb{C}), \{\cdot,\cdot\}),$$

let $\Omega^1(X,\mathbb{C})_{\{\cdot,\cdot\}}$ be the corresponding $(\mathbb{C}, C^\infty(X,\mathbb{C}))$-Lie algebra, and let $Y \subseteq X$ be an arbitrary stratum. Then the restriction map from $C^\infty(X,\mathbb{C})$ to $C^\infty(Y,\mathbb{C})$ is Poisson and hence induces a morphism

$$(C^\infty(X,\mathbb{C}), \Omega^1(X,\mathbb{C})_{\{\cdot,\cdot\}}) \to (C^\infty(Y,\mathbb{C}), \Omega^1(Y,\mathbb{C})_{\{\cdot,\cdot\}}) \tag{2.2}$$

of Lie-Rinehart algebras. Given a $(\mathbb{C}, C^\infty(X,\mathbb{C}))$-Lie subalgebra P of $\Omega^1(X,\mathbb{C})_{\{\cdot,\cdot\}}$, let P_Y be the $C^\infty(Y,\mathbb{C})$-submodule of $\Omega^1(Y,\mathbb{C})$ generated by the image of P under (2.2); it inherits the structure of a $(\mathbb{C}, C^\infty(Y,\mathbb{C}))$-Lie subalgebra of $\Omega^1(Y,\mathbb{C})_{\{\cdot,\cdot\}}$.

A $(\mathbb{C}, C^\infty(X,\mathbb{C}))$-Lie subalgebra P of $\Omega^1(X,\mathbb{C})_{\{\cdot,\cdot\}}$ will be said to be a *stratified (complex) polarization* for X if, for every stratum Y, under the isomorphism (2.1) of $(\mathbb{C}, C^\infty(Y,\mathbb{C}))$-Lie algebras (for the smooth symplectic manifold Y), the $(\mathbb{C}, C^\infty(Y,\mathbb{C}))$-Lie subalgebra P_Y of $\Omega^1(Y,\mathbb{C})_{\{\cdot,\cdot\}}$ is identified with the space of sections of a complex polarization in the usual sense. Given a stratified polarization P for X, a smooth complex function f in X, that is, a function which is defined on an open subset O of X and belongs to the induced smooth structure $C^\infty(O)$ will be said to be *polarized* provided it satisfies

$$\alpha(f) = 0, \quad \text{for every} \quad \alpha \in P.$$

Here and henceforth "smooth" refers to the structure algebras of functions $C^\infty(X)$ on X and $C^\infty(Y)$ on open subsets Y thereof; thus a "smooth function" in our sense is not necessarily an ordinary smooth function. In general polarized functions will exist at most locally, and we can talk about *polarized functions in X* or the *sheaf of germs of polarized functions*.

Infinitesimally, the notion of stratified polarization comes down to this: For $x \in X$, the Poisson 2-form $\pi_{\{\cdot,\cdot\}}$ passes to a 2-form Π_x on the cotangent space $\mathrm{T}^*_x X = \Omega^1(X) \otimes_{C^\infty(X)} \mathbb{R}_x$. Given a stratified polarization, when x lies in the stratum Y, the restriction map from

$$\mathrm{T}^{*\mathbb{C}}_x X = \mathrm{T}^*_x X \otimes \mathbb{C} = \Omega^1(X,\mathbb{C}) \otimes_{C^\infty(X,\mathbb{C})} \mathbb{C}_x$$

to $\mathrm{T}^{*\mathbb{C}}_x Y = \Omega^1(Y,\mathbb{C}) \otimes_{C^\infty(Y,\mathbb{C})} \mathbb{C}_x$ induces a surjective $\mathbb{C}$-linear map

$$P_x = P \otimes_{C^\infty(X,\mathbb{C})} \mathbb{C}_x \to P_x(Y) = P_Y \otimes_{C^\infty(Y,\mathbb{C})} \mathbb{C}_x,$$

and $P_x(Y)$ is a Lagrangian subspace of $\mathrm{T}^{*\mathbb{C}}_x Y$ with respect to the complexified symplectic structure $\pi^{\mathbb{C}}_x$ on $\mathrm{T}^{*\mathbb{C}}_x Y$.

When X is a smooth symplectic manifold, viewed as a stratified symplectic space with a single stratum, this notion of polarization manifestly boils down to the ordinary one. For a general stratified symplectic space, a stratified polarization encapsulates the *mutual positions of the polarizations on the strata*. Given a stratified symplectic space X, a stratified polarization P for X will be said to be a (positive) *stratified Kähler polarization* if, for every stratum Y, the image of P_Y under (2.1) is the space of sections of a (positive) Kähler polarization for Y in the usual sense. A stratified symplectic space with a stratified Kähler polarization will be said to be a *stratified Kähler space*. Since a Kähler polarization on a symplectic manifold determines a Kähler structure thereupon it is obvious that each stratum of a stratified Kähler space inherits a Kähler structure. Recall that a symplectic structure ω is said to be *compatible* with a complex structure J provided J is a symplectic operator so that ω defines a not necessarily positive Kähler structure.

PROPOSITION 2.3. *Given a real symplectic structure ω on a complex domain N of complex dimension n, let $\{\cdot,\cdot\}$ be the Poisson structure determined by ω. In terms of holomorphic coordinates $w_1, \ldots, w_n$, ω is compatible with the complex structure J if and only if $\{w_j, w_k\} = 0$ for $1 \le j,k \le n$. Under these circumstances, the Kähler polarization P (in our sense) is the free $C^\infty(N,\mathbb{C})$-submodule* *of* $\Omega^1(N,\mathbb{C})_{\{\cdot,\cdot\}}$ *generated by the differentials* $dw_1,\ldots,dw_n$. *For a smooth function* f *in* N, *the Cauchy-Riemann equations then amount to*

$$\{w_j,f\}=0,\quad 1\le j\le n.$$

PROOF. The differentials $dw_1,\ldots,dw_n,d\overline{w}_1,\ldots,d\overline{w}_n$ freely generate $\Omega^1(N,\mathbb{C})$ as a $C^\infty(N,\mathbb{C})$-module, and $\mathrm{Vect}(N,\mathbb{C})$ is, likewise, identifies the free $C^\infty(N,\mathbb{C})$-module generated by the vector fields $\frac{\partial}{\partial w_1},\ldots,\frac{\partial}{\partial w_n},\frac{\partial}{\partial \overline{w}_1},\ldots,\frac{\partial}{\partial \overline{w}_n}$. The symplectic structure ω on N is compatible with the complex structure J if and only if the isomorphism (2.1) of $(\mathbb{C},C^\infty(N,\mathbb{C}))$-Lie algebras from $\Omega^1(N,\mathbb{C})_{\{\cdot,\cdot\}}$ onto $\mathrm{Vect}(N,\mathbb{C})$ identifies the $C^\infty(N,\mathbb{C})$-linear span of the $dw_1,\ldots,dw_n$ with the $C^\infty(N,\mathbb{C})$-linear span of the $\frac{\partial}{\partial \overline{w}_1},\ldots,\frac{\partial}{\partial \overline{w}_n}$ (so that this span contains the hamiltonian vector fields $\{\cdot,w_j\}$ of the w_j's $(1\le j\le n)$ and therefore coincides with the span of these hamiltonian vector fields) and, likewise, the $C^\infty(N,\mathbb{C})$-linear span of the $d\overline{w}_1,\ldots,d\overline{w}_n$ with the $C^\infty(N,\mathbb{C})$-linear span of the $\frac{\partial}{\partial w_1},\ldots,\frac{\partial}{\partial w_n}$. The latter property, in turn, is equivalent to the vanishing of $\{w_j,w_k\}$ for $1\le j,k\le n$. See for example the discussion in **[95]** (5.4 p. 92). Under these circumstances, the Kähler polarization (in our sense) $P\subseteq\Omega^1(N,\mathbb{C})_{\{\cdot,\cdot\}}$ is generated by the differentials $dw_1,\ldots,dw_n$, and the *polarized* smooth complex functions f in N or, equivalently, the germs of *polarized* smooth complex functions are (classes of) functions f in N which satisfy $dw_j(f)=0$ for $1\le j\le n$, where $dw_j(\cdot)$ refers to the action of $dw_j\in\Omega^1(N,\mathbb{C})_{\{\cdot,\cdot\}}$ on $C^\infty(N,\mathbb{C})$ reproduced in Section 1 above. By construction, for every smooth complex function h in N, given f, we have $dh(f)=\{h,f\}$. Hence the *polarized* smooth complex functions f in N are those functions which satisfy the equations $\{w_j,f\}=0$ for $1\le j\le n$. However, for every coordinate function $w_j=q_j+ip_j$ and every smooth complex valued function f in N,

$$\{w_j,f\}=\sum\{w_j,\overline{w}_k\}\frac{\partial f}{\partial\overline{w}_k}.$$

Since ω is symplectic, the matrix $[\{w_j,\overline{w}_k\}]$ is invertible whence the Cauchy-Riemann equations amount to the vanishing of $\{w_j,f\}$ for $1\le j\le n$. □

Under the circumstances of Proposition 2.3, the holomorphic functions in N coincide with the smooth complex valued functions in N which are polarized in our sense.

Let X be a stratified Kähler space; we will say that a complex analytic structure on X is *compatible* with the stratified Kähler structure if (i) the stratification of X (as a stratified symplectic space) is a refinement of the complex analytic stratification, if (ii) every holomorphic function in X is smooth in X and if, (iii) the sheaf of germs of holomorphic functions in X is contained in its sheaf of germs of polarized functions. Given a stratified Kähler space X with a compatible complex analytic structure, for any stratum, endowed with the complex structure coming from the induced Kähler polarization on that stratum, the restriction map from X to that stratum is plainly complex analytic.

PROPOSITION 2.4. *Given a stratified Kähler space* X *together with a normal compatible complex analytic structure, if the stratification of* X *(as a stratified symplectic space) has an open and dense stratum, the sheaf of germs of holomorphic functions coincides with that of polarized functions.*

PROOF. The open and dense stratum is necessarily contained in the top complex analytic stratum. A continuous function in X which is holomorphic in the strata, in particular in the open and dense stratum, is necessarily itself holomorphic, since on a normal space Riemann's extension theorem holds. See e. g. WHITNEY'S book **[94]** or GRAUERT-REMMERT **[17]** for details. Consequently a polarized smooth complex function in X is necessarily holomorphic. □

We do not know whether for a general stratified Kähler space having a compatible complex analytic structure the sheaf of germs of holomorphic functions actually coincides with that of polarized functions, nor do we know whether a stratified Kähler polarization on a stratified symplectic space X necessarily determines a compatible complex analytic structure on X.

Borrowing some terminology from the language of sheaves, we will say that a smooth space X, with smooth structure $C^\infty(X)$, is *fine* provided for an arbitrary locally finite open covering $\mathcal{U} = \{U_\lambda\}$ of X, there is a partition of unity $\{u_\lambda\}$ subordinate to $\mathcal{U}$ (i. e. $\mathrm{supp}(u_\lambda) \subseteq U_\lambda$) with $u_\lambda \in C^\infty(X)$ for every λ; such a function u_λ will then be referred to as a *bump* function. The notion of fine space will help us avoid having to talk about the sheaf of germs of smooth functions.

THEOREM 2.5. *Let X be a fine stratified symplectic space, endowed with a complex analytic structure. Suppose that the symplectic stratification is a refinement of the complex analytic one and that germs of holomorphic functions belong to the stratified symplectic structure, that is, are smooth functions in X. Then the $C^\infty(X,\mathbb{C})$-submodule P of $\Omega^1(X,\mathbb{C})_{\{\cdot,\cdot\}}$ generated by differentials of the kind udf where f is holomorphic in X and u a bump function is a stratified Kähler polarization for X, necessarily compatible with the complex analytic structure, if and only if, for every pair f,h of holomorphic functions in X, the Poisson bracket $\{f,h\}$ vanishes.*

PROOF. Given differentials udf and vdh where u and v are bump functions and f and h holomorphic functions in X,

$$(*)\qquad [udf, vdh] = u\{f,v\}dh + v\{u,h\}df + uvd\{f,h\}.$$

Suppose that $\{f,h\}$ vanishes for any pair of holomorphic functions f and h in X. The identity $(*)$ implies that the $C^\infty(X,\mathbb{C})$-submodule P of $\Omega^1(X,\mathbb{C})_{\{\cdot,\cdot\}}$ is then closed under the Lie bracket. Let Y be a stratum of X. We must show that the $C^\infty(Y,\mathbb{C})$-submodule P_Y of $\Omega^1(Y,\mathbb{C})$ generated by the image of P under (2.2) passes under (2.1) (where Y plays the role of N in (2.1)) to the space of sections of an ordinary Kähler polarization on Y. Since the problem is local, we may suppose that X is the zero locus of a finite set $f_1,\dots,f_q$ of holomorphic functions on a polydisc U in some $\mathbb{C}^n$, with coordinates $w_1,\dots,w_n$. Restricted to the stratum Y, the holomorphic functions $w_1,\dots,w_n$ will not be independent but, since Y is a smooth complex submanifold of U, the $C^\infty(Y,\mathbb{C})$-submodule P_Y of $\Omega^1(Y,\mathbb{C})$ generated by the differentials $dw_1,\dots,dw_n$ coincides with the $C^\infty(Y,\mathbb{C})$-submodule P_Y of $\Omega^1(Y,\mathbb{C})$ generated by the holomorphic differentials in Y. More precisely, suitable holomorphic coordinate functions $\zeta_1,\dots,\zeta_\ell$ on Y may be written as holomorphic functions in the $w_1,\dots,w_n$, whence the $C^\infty(Y,\mathbb{C})$-submodule P_Y of $\Omega^1(Y,\mathbb{C})$ generated by the differentials $dw_1,\dots,dw_n$ coincides with that generated by the differentials $d\zeta_1,\dots,d\zeta_\ell$. By virtue of Proposition 2.3, the latter generate the Kähler polarization in our sense.

Conversely, when P is a stratified Kähler polarization for X, given holomorphic functions f and h in X, in view of Proposition 2.3, the bracket $\{f, h\}$ is zero on every stratum, that is, $\{f, h\}$ is zero. □

Under the circumstances of Theorem 2.5, the stratified Kähler polarization P is determined by the complex analytic structure; if, on the other hand, P *determines* the complex analytic structure in the sense that the sheaf of germs of polarized functions actually *coincides* with that of holomorphic functions, P will be said to be a *complex analytic stratified Kähler polarization*, and the corresponding space will be referred to as a *complex analytic stratified Kähler space*. A complex analytic stratified Kähler space which, as a complex analytic space, is normal will be said to be a *normal Kähler space*. Normal Kähler spaces constitute a particularly nice class of stratified Kähler spaces. All the examples of stratified Kähler spaces known to us are normal Kähler spaces.

The following is now a mere observation; yet we spell it out since it looks exceedingly attractive.

PROPOSITION 2.6. *A function f in a complex analytic stratified Kähler space X is holomorphic if and only if $\{w, f\} = 0$ for every holomorphic function w in X. Moreover, near any point p of X, the requisite complex analytic stratified Kähler polarization P has a finite set of $C^\omega(X, \mathbb{C})$-module generators $dw_1, \ldots, dw_n$ where $w_1, \ldots w_n$ are holomorphic functions in X defined near p, and a function f in X is holomorphic (near p) if and only if*

$$\{w_j, f\} = 0, \quad 1 \leq j \leq n. \qquad (2.6.1)$$ □

Under such circumstances, the equations (2.6.1) may be viewed as *Cauchy-Riemann* equations for the complex analytic stratified Kähler space X. *Thus the Cauchy-Riemann equations make sense even though X is not necessarily smooth.*

Henceforth every stratified symplectic space will be assumed to be fine. When X is a smooth symplectic manifold, viewed as a (fine) stratified symplectic space with a single stratum, Theorem 2.5 comes down to the usual characterization of a Kähler structure on X in terms of a Kähler polarization. However, to attack the question whether a stratified Kähler polarization on a general stratified symplectic space X necessarily determines a compatible complex analytic structure on X, one would have to develop a theory of obstructions to realizing germs of stratified Kähler spaces as complex analytic germs in a compatible fashion. In the smooth (i. e. manifold) case, there is no such realization problem since manifolds are locally modeled on affine spaces, and complex structures on affine spaces are handled by standard linear algebra.

3. The closure of a holomorphic nilpotent orbit

In this section we will show that the closure of a holomorphic nilpotent orbit inherits a normal complex analytic stratified Kähler structure. Moreover, we will give a complete classification of holomorphic nilpotent orbits, and we will explicitly describe the real and complex structures of the closure of any holomorphic nilpotent orbit.

3.1. REDUCTIVE LIE ALGEBRAS OF HERMITIAN TYPE. Following [**79**] (p. 54), we define a (semisimple) Lie algebra of *hermitian type* to be a pair $(\mathfrak{g}, z)$ which consists of a real semisimple Lie algebra $\mathfrak{g}$ with a Cartan decomposition $\mathfrak{g} = \mathfrak{k} \oplus \mathfrak{p}$ and a central element z of $\mathfrak{k}$, referred to as an *H-element*, such that $J_z = \mathrm{ad}(z)\big|_{\mathfrak{p}}$ is a (necessarily K-invariant) complex structure on $\mathfrak{p}$. Sometimes we will then say that $\mathfrak{g}$ is of hermitian type, without explicit reference to an H-element. Slightly more generally, a *reductive Lie algebra of hermitian type* is a reductive Lie algebra $\mathfrak{g}$ together with an element $z \in \mathfrak{g}$ whose constituent z' (say) in the semisimple part $[\mathfrak{g}, \mathfrak{g}]$ of $\mathfrak{g}$ is an H-element for $[\mathfrak{g}, \mathfrak{g}]$ [**79**] (p. 92). Given two semisimple (or reductive) Lie algebras $(\mathfrak{g}_1, z_1)$ and $(\mathfrak{g}_2, z_2)$ of hermitian type, a morphism $\phi\colon \mathfrak{g}_1 \to \mathfrak{g}_2$ of Lie algebras is said to be an *H_1-homomorphism* provided $\phi \circ \mathrm{ad}(z_1) = \mathrm{ad}(z_2) \circ \phi$. For a real semisimple Lie algebra $\mathfrak{g}$, with Cartan decomposition $\mathfrak{g} = \mathfrak{k} \oplus \mathfrak{p}$, we write G for its adjoint group and K for the (compact) connected subgroup of G with $\mathrm{Lie}(K) = \mathfrak{k}$; the requirement that $\mathfrak{g}$ be of hermitian type is equivalent to G/K being a (non-compact) hermitian symmetric space. Given $\mathfrak{g}$, the space of H-elements or, equivalently, the manifold of corresponding Cartan decompositions, may be identified with a disjoint union $(G/K^+) \cup (G/K^-)$ of two copies of the homogeneous space G/K in an obvious fashion, G/K^+ and G/K^- being the connected components containing z and $-z$, respectively. For example, for $\mathfrak{g} = \mathfrak{sp}(1, \mathbb{R}) = \mathfrak{sl}(2, \mathbb{R})$, G/K^+ and G/K^- are the two connected components of a standard hyperboloid.

Given a real semisimple Lie algebra $(\mathfrak{g}, z)$ of hermitian type, the underlying Lie algebra $\mathfrak{g}$ decomposes as

$$\mathfrak{g} = \mathfrak{g}_0 \oplus \mathfrak{g}_1 \oplus \ldots \oplus \mathfrak{g}_k \tag{3.1.1}$$

where $\mathfrak{g}_0$ is the maximal compact semisimple ideal and where $\mathfrak{g}_1, \ldots, \mathfrak{g}_k$ are non-compact and simple. For a non-compact simple Lie algebra with Cartan decomposition $\mathfrak{g} = \mathfrak{k} \oplus \mathfrak{p}$, the $\mathfrak{k}$-action on $\mathfrak{p}$ coming from the adjoint representation of $\mathfrak{g}$ is faithful and irreducible whence the center of $\mathfrak{k}$ is then at most one-dimensional; indeed $\mathfrak{g}$ has an H-element turning it into a Lie algebra of hermitian type if and only if the center of $\mathfrak{k}$ has dimension one. In view of E. Cartan's infinitesimal classification of irreducible hermitian symmetric spaces, the classical Lie algebras $\mathfrak{su}(p, q)$ $(A_n, n \geq 1$, where $n + 1 = p + q)$, $\mathfrak{so}(2, 2n - 1)$ $(B_n, n \geq 2)$, $\mathfrak{sp}(n, \mathbb{R})$ $(C_n, n \geq 2)$, $\mathfrak{so}(2, 2n - 2)$ $(D_{n,1}, n > 2)$, $\mathfrak{so}^*(2n)$ $(D_{n,2}, n > 2)$ together with the real forms $\mathfrak{e}_{6(-14)}$ and $\mathfrak{e}_{7(-25)}$ of the exceptional Lie algebras $\mathfrak{e}_6$ and $\mathfrak{e}_7$, respectively, constitute a complete list of simple Lie algebras of hermitian type. We refer to $\mathfrak{su}(p, q)$, $\mathfrak{sp}(n, \mathbb{R})$ and $\mathfrak{so}^*(2n)$ as the *standard cases*.

3.2. HOLOMORPHIC AND PSEUDOHOLOMORPHIC ORBITS. Let $(\mathfrak{g}, z)$ be a semisimple Lie algebra of hermitian type. Then, cf. [**79**], $\mathfrak{k} = \ker(\mathrm{ad}(z))$ and $\mathfrak{p} = [z, \mathfrak{g}]$, that is, the Cartan decomposition $\mathfrak{g} = \mathfrak{k} \oplus \mathfrak{p}$ is encapsulated in the choice of H-element z. As usual, let $\mathfrak{p}^{\mathbb{C}} = \mathfrak{p}^+ \oplus \mathfrak{p}^-$ be the decomposition of the complexification $\mathfrak{p}^{\mathbb{C}}$ of $\mathfrak{p}$ into $(+i)$- and $(-i)$-eigenspaces of J_z, respectively. Thus $\mathfrak{p}^+$ is the span of $w - iJ_z w$ $(= w - i[z, w])$ and $\mathfrak{p}^-$ that of $w + iJ_z w$ $(= w + i[z, w])$ where w runs through $\mathfrak{p}$.

The association $w \mapsto w - iJ_z w$ is an isomorphism of complex vector spaces from $\mathfrak{p}$, endowed with the complex structure J_z, onto $\mathfrak{p}^+$. We will refer to an adjoint orbit $\mathcal{O} \subseteq \mathfrak{g}$ having the property that the projection map from $\mathfrak{g}$ to $\mathfrak{p}$, restricted to $\mathcal{O}$, is a diffeomorphism onto its image, as a *pseudoholomorphic* orbit. In Lemma 3.3.6 below we shall show that a pseudoholomorphic nilpotent orbit maps (diffeomorphically) onto a smooth *complex* submanifold of $\mathfrak{p}$, and in (3.4) below we shall show that a non-nilpotent pseudoholomorphic orbit necessarily maps diffeomorphically onto $\mathfrak{p}$. Thus a pseudoholomorphic orbit $\mathcal{O}$ inherits a complex structure from the complex structure J_z on $\mathfrak{p}$, and this complex structure, combined with the Kostant-Kirillov-Souriau form on $\mathcal{O}$, viewed as a coadjoint orbit by means of (a positive multiple of) the Killing form, turns $\mathcal{O}$ into a (not necessarily positive) Kähler manifold. Indeed, a choice of (complex) basis $\zeta_1, \ldots, \zeta_k$ of $\mathfrak{p}^+$ yields holomorphic coordinate functions on $\mathfrak{p}^*$ which, together with their complex conjugates $\overline{\zeta}_1, \ldots, \overline{\zeta}_k$, may be used as coordinates for the smooth complex functions on $\mathfrak{p}^*$; since, by assumption, the composite of the injection of $\mathcal{O}$ into $\mathfrak{g}^* = \mathfrak{k}^* \oplus \mathfrak{p}^*$ (identified with $\mathfrak{g}$ via a suitable multiple of the Killing form, see below), combined with the projection to $\mathfrak{p}^*$, is a diffeomorphism onto its image, these functions yield coordinate functions on $\mathcal{O}$ which, with an abuse of notation, we still write as $\zeta_1, \ldots, \zeta_k$ and $\overline{\zeta}_1, \ldots, \overline{\zeta}_k$; we shall see below that every smooth complex function on $\mathcal{O}$ may be written as a function in these variables. Since the constituents $\mathfrak{p}^+$ and $\mathfrak{p}^-$ are abelian subalgebras of $\mathfrak{g}^{\mathbb{C}}$, cf. [**24**] (p. 313), [**79**] (p. 55), on $\mathcal{O}$ the Poisson-Lie brackets $\{\zeta_j, \zeta_{j'}\}$ vanish since the Poisson bracket on $\mathcal{O}$ is induced from the Lie bracket on $\mathfrak{g}$. Furthermore, since $[\mathfrak{p}, \mathfrak{p}] = \mathfrak{k}$, the Poisson-Lie brackets $\{\zeta_j, \overline{\zeta}_{j'}\}$ are given by the restriction to $\mathcal{O} \subset \mathfrak{k} \oplus \mathfrak{p}$ of the resulting complex functions on $\mathfrak{k}$ determined by the Lie bracket in $\mathfrak{k}$. In view of Proposition 2.3, the complex structure and the symplectic structure corresponding to the Poisson structure thus combine to a (not necessarily positive) Kähler structure on $\mathcal{O}$; this will be made precise in Theorem 3.7.1 below. For nilpotent orbits, we will make this somewhat roundabout reasoning more perspicuous in Theorem 3.3.11 via the classification given in Theorem 3.3.3 below.

We now choose a positive multiple of the Killing form. We will say that a pseudoholomorphic orbit $\mathcal{O}$ is *holomorphic* provided the resulting Kähler structure on $\mathcal{O}$ is positive. Any positive multiple of the Killing form will do; passing from one choice to another one amounts to rescaling the resulting hermitian form. The notion of holomorphic orbit relies on the choice of H-element z (while the notion of pseudoholomorphic orbit does not); given a holomorphic orbit $\mathcal{O}$, when z is replaced by $-z$, $-\mathcal{O}$ (and *not* $\mathcal{O}$) will be holomorphic with respect to $-z$. Sometimes we will refer to an orbit which is holomorphic with respect to $-z$ as an *antiholomorphic* orbit. The name "holomorphic" is intended to hint at the fact that the holomorphic discrete series representations arise from holomorphic quantization on integral semisimple holomorphic orbits but, beware, the requisite complex structure is not the one arising from projection to the symmetric constituent of the Cartan decomposition.

In particular, we can talk about *holomorphic nilpotent orbits*. A different definition of holomorphic nilpotent orbit may be found in [**89**] which, as we will show in Remark 3.3.13, is equivalent to ours.

Henceforth the *closure* is always understood to be taken in the *ordinary* (not Zariski) topology. Let $(\mathfrak{g}, z)$ be a semisimple Lie algebra of hermitian type. With reference to the decomposition (3.1.1), any (pseudo)holomorphic nilpotent orbit $\mathcal{O}$ in $\mathfrak{g}$ may be written as $\mathcal{O} = \mathcal{O}_1 \times \ldots \times \mathcal{O}_k$ where, for $1 \leq j \leq k$, each $\mathcal{O}_j$

is a (pseudo)holomorphic nilpotent orbit in $\mathfrak{g}_j$, with reference to the component z_j of z in $\mathfrak{g}_j$. Consequently, any orbit $\mathcal{O}'$ in the closure of $\mathcal{O}$ may be written as $\mathcal{O}' = \mathcal{O}'_1 \times \ldots \times \mathcal{O}'_k$ where, for $1 \leq j \leq k$, each $\mathcal{O}'_j$ is in the closure of $\mathcal{O}_j$. This reduces the study of general (pseudo)holomorphic nilpotent orbits (and of their closures) to that of (pseudo)holomorphic nilpotent orbits in simple hermitian Lie algebras.

We now spell out the main result of this section.

THEOREM 3.2.1. *Given a semisimple Lie algebra* $(\mathfrak{g}, z)$ *of hermitian type, for any holomorphic nilpotent orbit* $\mathcal{O}$*, the diffeomorphism from* $\mathcal{O}$ *onto its image in* $\mathfrak{p}$ *extends to a homeomorphism from the closure* $\overline{\mathcal{O}}$ *onto its image in* $\mathfrak{p}$*, this homeomorphism turns* $\overline{\mathcal{O}}$ *into a complex affine variety in such a way that,* $\mathfrak{g}$ *being suitably identified with its dual so that* $\overline{\mathcal{O}}$ *is identified with the closure of a coadjoint orbit, the following hold: The induced complex analytic structure on* $\overline{\mathcal{O}}$ *combines with the resulting Poisson structure to a normal complex analytic stratified Kähler structure, and the closure* $\overline{\mathcal{O}}$ *is a union of finitely many holomorphic nilpotent orbits. Furthermore, when* $\mathfrak{g}$ *is simple, a pseudoholomorphic nilpotent orbit* $\mathcal{O}$ *having the property that the diffeomorphism from* $\mathcal{O}$ *onto its image in* $\mathfrak{p}$ *extends to a homeomorphism from the closure* $\overline{\mathcal{O}}$ *onto its image in* $\mathfrak{p}$ *is necessarily holomorphic or antiholomorphic.*

Theorem 3.2.1 will be a consequence of Theorem 3.3.3 below save that the complex analytic stratified Kähler structure will be explained in Subsection 3.7 below. In the classical cases, the real and complex analytic structures of the closure $\overline{\mathcal{O}}$ of any holomorphic nilpotent orbit $\mathcal{O}$ will be elucidated in Theorems 3.5.4, 3.5.5, and 3.6.2 below. The normality will be explained in Theorems 3.5.5, 3.6.3, 5.3.3, 8.4.1, 8.4.2.

3.2.2. THE CASE OF $\mathfrak{sl}(2, \mathbb{R})$. This Lie algebra is (well known to be) of hermitian type, in the following fashion: The nilpotent elements $E = \begin{bmatrix} 0 & 1 \\ 0 & 0 \end{bmatrix}$ and $F = \begin{bmatrix} 0 & 0 \\ 1 & 0 \end{bmatrix}$ together with the semisimple one $H = \begin{bmatrix} 1 & 0 \\ 0 & -1 \end{bmatrix}$ span $\mathfrak{sl}(2, \mathbb{R})$, subject to the relations

$$[E, F] = H, \quad [H, E] = 2E, \quad [H, F] = -2F, \tag{3.2.2.1}$$

and $Z = \frac{1}{2}(E - F)$ is an H-element. The Cartan decomposition determined by the choice of Z has the form $\mathfrak{sl}(2, \mathbb{R}) = \mathfrak{k} \oplus \mathfrak{p}$ in such a way that $E - F$ spans $\mathfrak{k} \cong \mathbb{R}$ and $E + F$ and H span $\mathfrak{p}$. The induced complex structure J_Z on $\mathfrak{p}$ is given by

$$J_Z(E + F) = H, \quad J_Z(H) = -(E + F),$$

and $E + F$ is a basis of $\mathfrak{p}$, viewed as a complex vector space. Thus $(\mathfrak{sl}(2, \mathbb{R}), Z)$ is of hermitian type. More generally, for any $\ell \geq 1$, the product $(\mathfrak{sl}(2, \mathbb{R})^\ell, Z^\ell) = (\mathfrak{sl}(2, \mathbb{R})^\ell, Z, \ldots, Z)$ of ℓ copies of $(\mathfrak{sl}(2, \mathbb{R}), Z)$ is of hermitian type.

For intelligibility we reconcile the (well known) classification of orbits in $\mathfrak{sl}(2, \mathbb{R})$ with the notions of holomorphic and antiholomorphic orbit. The nilcone in $\mathfrak{sl}(2, \mathbb{R})$ consists of all $\alpha \in \mathfrak{sl}(2, \mathbb{R})$ with $\det(\alpha) = 0$ and decomposes into the disjoint union of the orbits $\mathcal{O}^+(0)$ and $\mathcal{O}^-(0)$ (say) of E and F, respectively, together with the orbit which consists merely of the origin. For $\varepsilon > 0$, the two-sheeted hyperboloid consisting of all $\alpha \in \mathfrak{sl}(2, \mathbb{R})$ with $\det(\alpha) = \varepsilon^2$ decomposes into the two elliptic

$\mathrm{SL}(2,\mathbb{R})$-orbits of $2\varepsilon Z$ and $-2\varepsilon Z$ which we denote by $\mathcal{O}^+(\varepsilon)$ and $\mathcal{O}^-(\varepsilon)$, respectively.

We will now show that the orbits $\mathcal{O}^+(\varepsilon)$ and $\mathcal{O}^-(\varepsilon)$ are holomorphic and antiholomorphic, respectively ($\varepsilon \geq 0$). To this end, let $b_0 = E - F$, $b_1 = E + F$, $b_2 = H$. For $\varepsilon > 0$, $\mathcal{O}^+(\varepsilon)$ and $\mathcal{O}^-(\varepsilon)$ are the G-orbits of εb_0 and $-\varepsilon b_0$, respectively and, when $x_0, x_1, x_2 \in \mathfrak{sl}(2,\mathbb{R})^*$ denote the coordinate functions in this basis, $\det(\alpha) = x_0^2 - x_1^2 - x_2^2$. Thus, for $\varepsilon \geq 0$,

$$\mathcal{O}^+(\varepsilon) = \{(x_0, x_1, x_2); x_1^2 + x_2^2 + \varepsilon^2 = x_0^2,\ x_0 > 0\}$$
$$\mathcal{O}^-(\varepsilon) = \{(x_0, x_1, x_2); x_1^2 + x_2^2 + \varepsilon^2 = x_0^2,\ x_0 < 0\}.$$

The Lie algebra $\mathfrak{sl}(2,\mathbb{R})$ may be identified with its dual via *any positive* multiple of the Killing form k ($k(\alpha,\beta) = 4\,\mathrm{trace}(\alpha\beta)$); the positivity is necessary in order for the isomorphism from $\mathfrak{p}$ to $\mathfrak{p}^*$ to preserve the natural orientations given by the complex structures. To obtain simple formulas, we will now identify $\mathfrak{sl}(2,\mathbb{R})$ and $\mathfrak{p}$ with their duals by means of the adjoint

$$\phi\colon \mathfrak{sl}(2,\mathbb{R}) \longrightarrow \mathfrak{sl}(2,\mathbb{R})^*, \quad \alpha \mapsto \phi_\alpha\colon \mathfrak{sl}(2,\mathbb{R}) \to \mathbb{R},$$

of the *trace* pairing (which is one fourth of the Killing form), that is, for $\alpha, \beta \in \mathfrak{sl}(2,\mathbb{R})$, $\phi_\alpha(\beta) = \mathrm{trace}(\alpha\beta)$. Then $\phi_{b_0} = -2x_0$, $\phi_{b_1} = 2x_1$, $\phi_{b_2} = 2x_2$, or, equivalently, when b_0, b_1, b_2 are viewed as linear functions on $\mathfrak{sl}(2,\mathbb{R})^*$,

$$\frac{b_0}{2} \circ \phi = -x_0, \ \frac{b_1}{2} \circ \phi = x_1, \ \frac{b_2}{2} \circ \phi = x_2.$$

The Poisson structure on $\mathfrak{sl}(2,\mathbb{R})$ induced via ϕ is determined by (3.2.2.1) and is therefore given by the formulas

$$\{x_1, x_2\} = x_0,\ \{x_0, x_2\} = x_1,\ \{x_0, x_1\} = -x_2. \tag{3.2.2.2}$$

The orthogonal projection from $\mathfrak{sl}(2,\mathbb{R})$ to the plane $\mathfrak{p}$, restricted to $\mathcal{O}^+(\varepsilon)$ or to $\mathcal{O}^-(\varepsilon)$ ($\varepsilon \geq 0$), is plainly a diffeomorphism onto its image in $\mathfrak{p}$ which is onto $\mathfrak{p}$ for $\varepsilon > 0$; for $\varepsilon = 0$, this diffeomorphism extends to a homeomorphism from $\overline{\mathcal{O}^+(0)}$ as well as from $\overline{\mathcal{O}^-(0)}$ onto $\mathfrak{p}$. For $\varepsilon \geq 0$, we now parametrize the orbits $\mathcal{O}^\pm(\varepsilon)$ by the plane $\mathfrak{p}$. For $\varepsilon > 0$, we take x_1 and x_2 as coordinate functions for the plane $\mathfrak{p}$ while, for $\varepsilon = 0$, the requisite (non-standard) smooth structure C^∞ on $\mathfrak{p}$ is the algebra of smooth functions in the variables x_0, x_1, x_2, subject to the relation $x_1^2 + x_2^2 = x_0^2$. The Poisson structures induced on $\mathfrak{p}$ from the Poisson structures on the orbits via the parametrizations are now given by

$$\{x_1, x_2\} = x_0 = \begin{cases} \sqrt{\varepsilon^2 + (x_1^2 + x_2^2)} & \text{for } \ \mathcal{O}^+(\varepsilon) \\ -\sqrt{\varepsilon^2 + (x_1^2 + x_2^2)} & \text{for } \ \mathcal{O}^-(\varepsilon) \end{cases}$$

when $\varepsilon > 0$ and by (3.2.2.2), subject to the relation $x_1^2 + x_2^2 = x_0^2$, when $\varepsilon = 0$. In the latter case, the Poisson structure extends the symplectic Poisson structure on $\mathcal{O}^+(0)$ (as well as that on $\mathcal{O}^-(0)$) to one on $\overline{\mathcal{O}^+(0)} \cong \mathfrak{p}$ (or $\overline{\mathcal{O}^-(0)} \cong \mathfrak{p}$) (cf. [**33, 39, 40**]). The induced complex structure J_Z on $\mathfrak{p}$ is standard, with holomorphic coordinate $\zeta = x_1 + ix_2$. Consequently, when $\mathfrak{p}$ is used to parametrize $\mathcal{O}^+(\varepsilon)$, the resulting hermitian form may be written as $\frac{d\zeta d\overline{\zeta}}{\sqrt{\varepsilon^2 + \zeta\overline{\zeta}}}$ and, when $\mathfrak{p}$ is used to parametrize $\mathcal{O}^-(\varepsilon)$, this form may accordingly be written as $-\frac{d\zeta d\overline{\zeta}}{\sqrt{\varepsilon^2 + \zeta\overline{\zeta}}}$. Thus, the orbits $\mathcal{O}^+(\varepsilon)$ are holomorphic and the orbits $\mathcal{O}^-(\varepsilon)$ are antiholomorphic, whatever

$\varepsilon \geq 0$, and the Poisson and complex analytic structures turn $\overline{\mathcal{O}^+(0)}$ into a normal positive Kähler space and, likewise, $\mathcal{O}^-(0)$ into a normal "negative" Kähler space. Moreover, for $\varepsilon > 0$, the Kähler manifold $\mathcal{O}^+(\varepsilon)$ *contracts* into the normal Kähler space $\overline{\mathcal{O}^+(0)}$ and, likewise, $\mathcal{O}^-(\varepsilon)$ contracts into $\overline{\mathcal{O}^-(0)}$. Hence, ε being viewed as a deformation parameter, the *singular* normal Kähler structure on $\overline{\mathcal{O}^+(0)}$ *deforms into an ordinary Kähler structure*; under this deformation, the non-standard Poisson structure "resolves" into an ordinary smooth symplectic Poisson structure while the complex analytic structure is *not* deformed. In Section 9 we will recall the notion of contraction and show how this relationship between $\overline{\mathcal{O}^+(0)}$ and the $\mathcal{O}^+(\varepsilon)$'s ($\varepsilon > 0$) extends to arbitrary holomorphic nilpotent orbits.

For $\varepsilon > 0$, the complex structure J_Z (say) on $\mathcal{O}^+(\varepsilon)$ and $\mathcal{O}^-(\varepsilon)$ induced from the complex structure J_Z on $\mathfrak{p}$ via the projection to $\mathfrak{p}$ is *not* the hermitian symmetric space structure. In order to see this, we note that, for $\varepsilon > 0$, the stereographic projection

$$(x_0, x_1, x_2) \longmapsto (y_1, y_2) = \left(\frac{x_1}{1 + \frac{x_0}{\varepsilon}}, \frac{x_2}{1 + \frac{x_0}{\varepsilon}}\right)$$

identifies each $\mathcal{O}^+(\varepsilon)$ with the generalized Poincaré disc $D_\varepsilon = \{(y_1, y_2) : y_1^2 + y_2^2 < \varepsilon\}$ in the (y_1, y_2)-plane, generalized in the sense that, in the literature, for $\varepsilon = 1$, this is the familiar Poincaré disc; see e. g. (3.7.5) in [**5**]. A straightforward calculation gives $\{y_1, y_2\} = \frac{\varepsilon}{4}\left(1 - \frac{r^2}{\varepsilon^2}\right)^2$ where $r^2 = y_1^2 + y_2^2$ whence the induced symplectic structure ω on D_ε may be written as $\omega = \frac{1}{\varepsilon}\frac{4}{\left(1-\frac{r^2}{\varepsilon^2}\right)^2} dy_1 \wedge dy_2$. This is the Kähler form of the hermitian metric $\frac{1}{\varepsilon}\frac{4}{\left(1-\frac{r^2}{\varepsilon^2}\right)^2} d\sigma d\overline{\sigma}$ on D_ε with holomorphic coordinate $\sigma = y_1 + iy_2$ to which the hermitian symmetric space hermitian form on $\mathcal{O}^+(\varepsilon)$ passes under the stereographic projection. Thus the latter is a symplectomorphism but it is *not* compatible with the complex structures when $\mathcal{O}^+(\varepsilon)$ is endowed with the complex structure J_Z since then the coordinate change $(x_1, x_2) \mapsto (y_1, y_2)$ is not holomorphic. Thus, when endowed with the Kähler structure determined by J_Z, $\mathcal{O}^+(1)$ is only *symplectically* a model for the Siegel space (the space of complex structures on the plane compatible with its standard symplectic structure).

REMARK. The hermitian symmetric space Kähler metric on $\mathcal{O}^+(\varepsilon)$ (equivalently: on D_ε, with the metric $\frac{1}{\varepsilon}\frac{4}{\left(1-\frac{r^2}{\varepsilon^2}\right)^2} d\sigma d\overline{\sigma}$) has constant negative curvature equal to $-\frac{1}{\varepsilon^2}$ while the hermitian form $\frac{d\zeta d\overline{\zeta}}{\sqrt{\varepsilon^2 + \zeta\overline{\zeta}}}$ has strictly negative curvature which is constant *only* along K-orbits, $K = \mathrm{U}(1)$ being the maximal compact subgroup of $\mathrm{SL}(2, \mathbb{R})$ determined by the choice of J_Z. Indeed, a calculation shows that, for $\varepsilon \geq 0$, $(\mathcal{O}^+(\varepsilon), J_Z)$ has curvature equal to $-\frac{\varepsilon^2}{(\varepsilon^2 + \zeta\overline{\zeta})^{\frac{5}{2}}}$; notice that $\zeta \neq 0$ when $\varepsilon = 0$. This shows again that, for $\varepsilon > 0$, J_Z does not yield the hermitian symmetric space structure on $\mathcal{O}^+(\varepsilon)$. Moreover, this formula entails that, in particular, $\mathcal{O}^+(0)$ (as well as $\mathcal{O}^-(0)$) is flat (has zero curvature).

3.3. CLASSIFICATION OF PSEUDOHOLOMORPHIC NILPOTENT ORBITS. Let $\mathfrak{g}$ be a real Lie algebra. As usual, we will refer to a triple (e, f, h) of elements of $\mathfrak{g}$ as an $\mathfrak{sl}(2)$-*triple* in $\mathfrak{g}$ provided $[h, e] = 2e$, $[h, f] = -2f$, $[e, f] = h$. An $\mathfrak{sl}(2)$-triple (e, f, h) plainly determines and is determined by a homomorphism $\kappa\colon \mathfrak{sl}(2, \mathbb{R}) \to \mathfrak{g}$ of real Lie algebras via $\kappa(E) = e$, $\kappa(F) = f$, $\kappa(H) = h$. Suppose in addition that $\mathfrak{g}$ is semisimple, with Cartan involution ϑ and Cartan decomposition $\mathfrak{g} = \mathfrak{k} \oplus \mathfrak{p}$. We will say that an $\mathfrak{sl}(2)$-triple (e, f, h) in $\mathfrak{g}$ is *invariant* provided $e - f \in \mathfrak{k}$ and $e + f, h \in \mathfrak{p}$. An $\mathfrak{sl}(2)$-triple (e, f, h) in $\mathfrak{g}$ is plainly invariant if and only if the corresponding Lie algebra homomorphism κ is compatible with the Cartan involutions.

Let $(\mathfrak{g}, z)$ be a simple Lie algebra of Hermitian type, with Cartan decomposition $\mathfrak{g} = \mathfrak{k} \oplus \mathfrak{p}$. We will say that an invariant $\mathfrak{sl}(2)$-triple is *holomorphic* or an H_1-triple provided the Lie algebra homomorphism from $\mathfrak{sl}(2, \mathbb{R})$ to $\mathfrak{g}$ it determines is an H_1-homomorphism. An invariant $\mathfrak{sl}(2)$-triple (e, f, h) is plainly an H_1-triple if and only if

$$(3.3.1) \qquad J_z(e+f)(= [z, e+f]) = h, \quad J_z(h)(= [z, h]) = -(e+f).$$

We will deduce the classification of holomorphic nilpotent orbits in $\mathfrak{g}$ from the relative root system. We will systematically denote the split rank, also called real rank (the dimension of a maximal abelian subalgebra of $\mathfrak{p}$) by r. The following result is a version of an observation spelled out as Remark 3 on p. 62 of [**79**] and we therefore label it as a Proposition. This observation, in turn, is a consequence of the existence of r strongly orthogonal (absolute) roots relative to the complexification of a compact Cartan subalgebra (referred to as well as *compactly embedded* Cartan subalgebra in the literature), cf. Lemma II.4.3 on p. 60 of [**79**].

PROPOSITION 3.3.2. *Given a simple Lie algebra $(\mathfrak{g}, z)$ of Hermitian type and split rank r, with Cartan decomposition $\mathfrak{g} = \mathfrak{k} \oplus \mathfrak{p}$, a choice of maximal abelian subalgebra $\mathfrak{a}_{\mathfrak{p}}$ of $\mathfrak{p}$ determines a system (e_1, f_1, h_1), ... ,(e_r, f_r, h_r) of H_1-triples which combine to an H_1-embedding of $(\mathfrak{sl}(2, \mathbb{R})^r, Z^r)$ into $(\mathfrak{g}, z)$ in such a way that the following hold.*
1) *The elements $h_1, \ldots, h_r$ constitute a basis of $\mathfrak{a}_{\mathfrak{p}}$.*
2) *With the notation $(\xi_1, \ldots, \xi_r)$ for the basis of $\mathfrak{a}_{\mathfrak{p}}^*$ dual to $h_1, \ldots, h_r$, the relative root system $S = S(\mathfrak{g}, \mathfrak{a}_{\mathfrak{p}}) \subseteq \mathfrak{a}_{\mathfrak{p}}^*$ determined by $\mathfrak{a}_{\mathfrak{p}}$ is given by*

$$S = \begin{cases} \{\pm\xi_j \pm \xi_k \, (j \neq k), \ \pm 2\xi_j\} & \text{type } (\mathrm{C}_r), \quad \text{or} \\ \{\pm\xi_j \pm \xi_k \, (j \neq k), \ \pm 2\xi_j, \pm\xi_j\} & \text{type } (\mathrm{BC}_r). \end{cases}$$

3) *The H-element z of $\mathfrak{g}$ equals $\frac{1}{2}(e_1 - f_1 + \cdots + e_r - f_r)$ if and only if S is of type (C_r).*
4) *The H_1-embedding of $(\mathfrak{sl}(2, \mathbb{R})^r, Z^r)$ into $(\mathfrak{g}, z)$ extends to an H_1-embedding of $\mathfrak{sp}(r, \mathbb{R})$ into $\mathfrak{g}$, $\mathfrak{sp}(r, \mathbb{R})$ being endowed with the appropriate H-element, and this embedding identifies the two H-elements if and only if S is of type (C_r).*

We will say that a simple Lie algebra of hermitian type is *regular* provided its relative root system is of type (C_r), and we will say that it is *non-regular* otherwise. This notion of regularity fits with that of regularity of pre-homogeneous spaces to be examined in Section 7 below. For the corresponding symmetric domain, regularity amounts to the property of being a tube domain.

PROOF. To adjust the notation on p. 110 of [**79**] to ours, recall that in [**79**] (p. 92 and p. 109) certain root vectors $X_k, Y_k, H_k \in \mathfrak{g}_{\mathbb{C}}$ $(1 \leq k \leq r)$ are given satisfying the equations

$$[H_k, X_k] = -2Y_k, \quad [H_k, Y_k] = 2X_k, \quad [X_k, Y_k] = 2H_k.$$

Then the triples (e_k, f_k, h_k) given by

$$e_k = \frac{1}{2}(Y_k + H_k), \quad f_k = \frac{1}{2}(Y_k - H_k), \quad h_k = X_k$$

(cf. III.1.6. on p. 91 and III.2.9 on p. 97 of [**79**]) will have the properties asserted in 1) and 2).

The statement 3) is a consequence of Corollary III.1.6 on p. 94, combined with Remark 1 in III.8 on p. 150 of [**79**]. To justify statement 4) we observe that, in the regular case, appropriate root vectors for the relative root system S span a copy of $\mathfrak{sp}(r, \mathbb{R})$ in $\mathfrak{g}$. In the non-regular case, the relative root system $S = \{\pm\xi_j \pm \xi_k\,(j \neq k),\ \pm 2\xi_j, \pm\xi_j\}$ contains the subsystem $S' = \{\pm\xi_j \pm \xi_k\,(j \neq k),\ \pm 2\xi_j\}$ and, likewise, appropriate root vectors for the latter span a copy of $\mathfrak{sp}(r, \mathbb{R})$ in $\mathfrak{g}$. □

REMARK. For intelligibility, we list the various types of the relative root systems that occur here, cf. [**79**] p. 115:
$\mathfrak{su}(q,q)$: (C_q); $\mathfrak{su}(p,q)$, $p > q$: (BC_q); $\mathfrak{so}(2,q)$: (C_2) $(q \geq 3)$; $\mathfrak{sp}(n,\mathbb{R})$: (C_n);
$\mathfrak{so}^*(2n)$: $\begin{cases} (\mathrm{C}_{\frac{n}{2}}), & n \text{ even} \\ (\mathrm{BC}_{[\frac{n}{2}]}), & n > 2 \text{ odd} \end{cases}$; $\mathfrak{e}_{6(-14)}$: (BC_2); $\mathfrak{e}_{7(-25)}$: (C_3).

Given a simple Lie algebra $(\mathfrak{g}, z)$, with adjoint group G, for $0 \leq t, u \leq r$, $t+u \leq r$, let $e_{t,u} = e_1 + \cdots + e_t - e_{t+1} - \cdots - e_{t+u} \in \mathfrak{g}$ and let $\mathcal{O}_{t,u} = Ge_{t,u} \subseteq \mathfrak{g}$; this is plainly a nilpotent orbit in $\mathfrak{g}$. We will refer to (t,u) as the *type* of $\mathcal{O}_{t,u}$, to $t+u$ as its *rank*, and to $t-u$ as its *signature*. For later reference, we will write $f_{t,u} = f_1 + \cdots + f_t - f_{t+1} - \cdots - f_{t+u} \in \mathfrak{g}$ $(0 \leq t, u \leq r,\ t+u \leq r)$.

THEOREM 3.3.3. *For a simple Lie algebra $(\mathfrak{g}, z)$ of hermitian type and split rank r, the $\frac{(r+1)(r+2)}{2}$ nilpotent orbits $\mathcal{O}_{t,u}$ $(0 \leq t, u \leq r,\ t+u \leq r)$ constitute a complete list of the pseudoholomorphic nilpotent orbits. Given such an orbit $\mathcal{O}_{t,u}$, an orbit $\mathcal{O}'$ distinct from $\mathcal{O}_{t,u}$ is in the closure of $\mathcal{O}_{t,u}$ if and only if $\mathcal{O}' = \mathcal{O}_{t',u'}$ for some $t' \leq t$ and $u' \leq u$ with $t'+u' < t+u$. In particular, the orbits $\mathcal{O}_{t,0}$ $(t \leq r)$ and $\mathcal{O}_{0,u}$ $(u \leq r)$ are precisely the holomorphic and antiholomorphic ones, respectively.*

ADDENDUM. *For a nilpotent orbit $\mathcal{O}$ of a simple Lie algebra $(\mathfrak{g}, z)$ of hermitian type, the property of being holomorphic, antiholomorphic, or pseudoholomorphic does not depend on the choice of H-element z. More precisely: Once an H-element z has been chosen, when $\mathcal{O}$ is holomorphic or antiholomorphic with respect to z, it will be so with respect to any $\widetilde{z}$ in the space G/K of H-elements; when $\mathcal{O}$ is pseudoholomorphic with respect to z, it will be so with respect to any $\widetilde{z}$ in the space $(G/K^+) \cup (G/K^-)$.*

The proof requires some preparation: Following the procedure in III.4 of [**79**] (p. 110), we choose the basis $\Delta = \{\gamma_1, \ldots, \gamma_r\}$ of S where $\gamma_j = \xi_j - \xi_{j+1}$ for $1 \leq j \leq r-1$ and where $\gamma_r = \xi_r$ in the regular case and $\gamma_r = 2\xi_r$ in the non-regular case. Let $S = S^+ \cup S^-$ be the resulting decomposition into positive and negative relative roots, so that

$$S^+ = \begin{cases} \{\xi_j \pm \xi_k\,(j<k),\ 2\xi_j\} & \text{regular case} \\ \{\xi_j \pm \xi_k\,(j<k),\ 2\xi_j, \xi_j\} & \text{non-regular case.} \end{cases}$$

For $\lambda \in S$, let $\mathfrak{g}^\lambda$ be its root space and, for any $S' \subseteq S$, write $\mathfrak{g}^{S'} = \sum_{\lambda \in S'} \mathfrak{g}^\lambda$. Thus $\mathfrak{g}$ decomposes as $\mathfrak{g} = \mathfrak{l} \oplus \mathfrak{g}^S$ where $\mathfrak{l} = \mathfrak{m} + \mathfrak{a}_\mathfrak{p}$ is the centralizer of $\mathfrak{a}_\mathfrak{p}$ in $\mathfrak{g}$, $\mathfrak{m}$

being the centralizer of $\mathfrak{a}_\mathfrak{p}$ in $\mathfrak{k}$. With the notation $\mathfrak{n}^+ = \mathfrak{g}^{S^+}$, the resulting Iwasawa decomposition has the form $\mathfrak{g} = \mathfrak{k} + \mathfrak{a}_\mathfrak{p} + \mathfrak{n}^+$; accordingly, we write $G = KA_\mathfrak{p}N$ for the corresponding global Iwasawa decomposition—in particular, this fixes the nilpotent group N—and, for later reference, we write $\mathfrak{n}^- = \mathfrak{g}^{S^-}$.

For $k = 1, \dots, r$, let

$$e^k = e_1 + \cdots + e_k,\ f^k = f_1 + \cdots + f_k,\ h^k = h_1 + \cdots + h_k;$$

this yields the H_1-triple (e^k, f^k, h^k) and we write $\kappa_k \colon \mathfrak{sl}(2,\mathbb{R}) \to \mathfrak{g}$ for the resulting H_1-homomorphism. Here and below the ordering of the H_1-triples (e_k, f_k, h_k) $(1 \leq k \leq r)$ does not matter. For $k = 1, \dots, r$, the parabolic subalgebra $\mathfrak{q}_k$ of $\mathfrak{g}$ (written as $\mathfrak{b}_\kappa$ in [**79**] III.2 (p. 95)) which is determined by (e^k, f^k, h^k) or, what amounts to the same, by κ_k, has Levi decomposition $\mathfrak{q}_k = \mathfrak{l}_k \oplus \mathfrak{n}_k^+$ where $\mathfrak{l}_k$ is the (reductive) centralizer of h^k in $\mathfrak{g}$, and

$$\mathfrak{g} = \mathfrak{n}_k^- \oplus \mathfrak{l}_k \oplus \mathfrak{n}_k^+. \tag{3.3.4(k)}$$

Let $\mathfrak{n}^+(0) = \{x \in \mathfrak{n}^+; [h^r, x] = 0\}$, $\mathfrak{n}^-(0) = \{x \in \mathfrak{n}^-; [h^r, x] = 0\}$, for $j = 1, 2$, let

$$\mathfrak{n}(j) = \{x \in \mathfrak{n}^+; [h^r, x] = jx\}, \quad \mathfrak{n}(-j) = \{x \in \mathfrak{n}^-; [h^r, x] = -jx\}$$

and, more generally, given $1 \leq k \leq r$, let

$$\mathfrak{n}_k(j) = \{x \in \mathfrak{n}^+; [h^k, x] = jx\},\ \mathfrak{n}_k(-j) = \{x \in \mathfrak{n}^-; [h^k, x] = -jx\}.$$

We note that $\mathfrak{n}_k(1)$ and $\mathfrak{n}_k(2)$ correspond to V_κ and U_κ in III.2 of [**79**] (p. 95 ff.), respectively. The following is well known (and easy to prove).

LEMMA 3.3.5. *As a vector space, the nilpotent Lie algebra $\mathfrak{n}^+$ decomposes as*

$$\mathfrak{n}^+ = \mathfrak{n}^+(0) \oplus \mathfrak{n}(1) \oplus \mathfrak{n}(2),$$

and the decomposition yields a graded Lie algebra ("graded" being understood in the obvious naive sense) whence, in particular, $\mathfrak{n}(2)$ is abelian. Furthermore, in the regular case, $\mathfrak{n}(1)$ is zero while, in the non-regular case, $\mathfrak{n}_r^+$ (which, as a vector space, is then the direct sum $\mathfrak{n}(1) \oplus \mathfrak{n}(2)$, cf. (3.3.4(k)) for $k = 3$ and what is said above,) may be written as a central Lie algebra extension

$$0 \longrightarrow \mathfrak{n}(2) \longrightarrow \mathfrak{n}_r^+ \longrightarrow \mathfrak{n}(1) \longrightarrow 0$$

with abelian quotient, i. e. as a Heisenberg algebra. More generally, for $1 \leq k \leq r$, as a vector space, $\mathfrak{n}_k^+$ decomposes as $\mathfrak{n}_k^+ = \mathfrak{n}_k(1) \oplus \mathfrak{n}_k(2)$. □

To spell out the next lemma we recall that the linear K-action on $\mathfrak{p}$, viewed as a complex vector space via the complex structure J_z, extends canonically to a linear $K^{\mathbb{C}}$-action on this vector space.

LEMMA 3.3.6. *Each nilpotent orbit $\mathcal{O}_{t,u}(= Ge_{t,u})$ (where $0 \leq t, u \leq r$, $t + u \leq r$) is pseudoholomorphic, that is, the projection from $\mathcal{O}_{t,u}$ to $\mathfrak{p}$ is a diffeomorphism onto its image in $\mathfrak{p}$, and the image of $\mathcal{O}_{t,u}$ in $\mathfrak{p}$ is the $K^{\mathbb{C}}$-orbit of $\frac{1}{2}(e_{t,u} + f_{t,u})$.*

PROOF. Write $\mathfrak{n}_\mathfrak{p}^+ \subseteq \mathfrak{n}(2)$ and $\mathfrak{n}_\mathfrak{p}^- \subseteq \mathfrak{n}(-2)$ for the span of $e_1, e_2, \dots, e_r$ and $f_1, f_2, \dots, f_r$, respectively. We claim that $G\mathfrak{n}_\mathfrak{p}^+ = K\mathfrak{n}_\mathfrak{p}^+$. Indeed, $G\mathfrak{n}_\mathfrak{p}^+$ is the union of the orbits $\mathcal{O}_{t,u}$ $(0 \le t, u \le r,\ t+u \le r)$. By Lemma 3.3.5, $[\mathfrak{n}^+, \mathfrak{n}(2)] \subseteq \mathfrak{n}(2)$; hence $N\mathfrak{n}_\mathfrak{p}^+ = \mathfrak{n}_\mathfrak{p}^+$ since N acts trivially on $\mathfrak{n}^+(2)$. Furthermore, $\mathfrak{n}_\mathfrak{p}^+$ is closed under the $A_\mathfrak{p}$-action. Thus, by virtue of the Iwasawa decomposition $G = KA_\mathfrak{p}N$ of G, $G\mathfrak{n}_\mathfrak{p}^+ = K\mathfrak{n}_\mathfrak{p}^+$ as asserted. Consequently the orbit $\mathcal{O}_{t,u} = G(e_1 + \dots + e_t - e_{t+1} - \dots - e_{t+u})$ decomposes into the K-orbits of the elements

$$e_{t,u,\lambda} = \lambda_1 e_1 + \dots + \lambda_t e_t - \lambda_{t+1} e_{t+1} - \dots - \lambda_{t+u} e_{t+u},$$

where $\lambda = (\lambda_1, \dots, \lambda_{t+u}) \in \mathbb{R}^{t+u}$ with $\lambda_j > 0$ $(1 \le j \le t+u)$. In view of the action of the relative Weyl group, which contains a copy of the symmetric group on r letters, we normalize λ by requiring $\lambda_1 \le \lambda_2 \le \dots \le \lambda_t$ and $\lambda_{t+1} \le \dots \le \lambda_{t+u}$. Likewise the orbit $G(f_1 + \dots + f_t - f_{t+1} - \dots - f_{t+u})$ decomposes into the K-orbits of the elements

$$f_{t,u,\mu} = \mu_1 f_1 + \dots + \mu_t f_t - \mu_{t+1} f_{t+1} - \dots - \mu_{t+u} f_{t+u}$$

where $\mu = (\mu_1, \dots, \mu_t, \mu_{t+1}, \dots, \mu_{t+u}) \in \mathbb{R}^{t+u}$ with $\mu_j > 0$ $(1 \le j \le t+u)$.

The image in $\mathfrak{p}$ of $e_{t,u,\lambda}$ under the orthogonal projection to $\mathfrak{p}$ is the element $\frac{1}{2}(e_{t,u,\lambda} + f_{t,u,\lambda}) \in \mathfrak{p}$. Using the (known) structure of the stabilizer of any nilpotent element in a semisimple Lie algebra, cf. e. g. Theorem 1.7 in [**25**], we conclude that the stabilizer $Z_K(\frac{1}{2}(e_{t,u,\lambda} + f_{t,u,\lambda}))$ in K of $\frac{1}{2}(e_{t,u,\lambda} + f_{t,u,\lambda}) \in \mathfrak{p}$ coincides with the stabilizer $Z_K(e_{t,u,\lambda})$ in K of $e_{t,u,\lambda}$. More precisely: The nilpotent element $e = e_{t,u,\lambda}$ belongs to the invariant $\mathfrak{sl}(2)$-triple (e, f, h) where

$$f = \frac{1}{\lambda_1} f_1 + \dots + \frac{1}{\lambda_t} f_t - \frac{1}{\lambda_{t+1}} f_{t+1} - \dots - \frac{1}{\lambda_{t+u}} f_{t+u}$$

and $h = h_1 + \dots + h_t + h_{t+1} + \dots + h_{t+u}$; we note that (e, f, h) is an H_1-triple if and only if $u = 0$ and if $\lambda_1 = \dots = \lambda_{t+u} = 1$. The stabilizer $Z_G(e)$ of e in the adjoint group G of $\mathfrak{g}$ is contained in the Jacobson-Morozow parabolic subgroup $Q = UL$ (say) associated with (e, f, h) (where U is the unipotent radical and L the Levi subgroup), and the reductive constituent of $Z_G(e)$ is contained in the (reductive) stabilizer $Z_G(\mathfrak{s})$ of the entire triple (where $\mathfrak{s}$ refers to the copy of $\mathfrak{sl}(2, \mathbb{R})$ in $\mathfrak{g}$ generated by e, f, h) whence $Z_K(e) = Z_G(e) \cap K$ is contained in $Z_G(\mathfrak{s})$, in fact, equals $Z_G(\mathfrak{s}) \cap K$; the notation in [**25**] is (X, Y, H) for our (e, f, h). Noting that, with reference to the Cartan involution ϑ on $\mathfrak{g}$, $f_{t,u,\lambda} = -\vartheta(e_{t,u,\lambda}) = -\vartheta(e)$, we must show that the inclusion $Z_K(e) \subseteq Z_K(\frac{1}{2}(e - \vartheta e))$ is the identity. In order to justify this assertion, we note first that, when $u = 0$ and $\lambda_1 = \dots = \lambda_t = 1$, we have $e = e_1 + \dots + e_t = e^t$, $f = f_1 + \dots + f_t = f^t$, and $J_z(e+f) = h = h_1 + \dots + h_t = h^t$; this implies that $Z_K(e+f) = Z_K(h) = K \cap L_t$ (since J_z commutes with the adjoint action of K on $\mathfrak{g}$) where L_t refers to the Levi subgroup of the parabolic subgroup $Q_t = U_t L_t$ associated with the H_1-triple (e^t, f^t, h^t). Thus $Z_K(e+f)$ now coincides with the stabilizer in K of h. The latter, in turn, equals the stabilizer $Z_K(e)$ of e in K; since the factor $\frac{1}{2}$ is irrelevant, we conclude that $Z_K(\frac{1}{2}(e+f))$ is equal to $Z_K(e)$ in this special case. Next we consider the case where $e = e_1 + \dots + e_t - e_{t+1} - \dots - e_{t+u}$ and $f = f_1 + \dots + f_t - f_{t+1} - \dots - f_{t+u}$. Noting that e and f are conjugate to $e_1 + \dots + e_t + e_{t+1} + \dots + e_{t+u}$ and $f_1 + \dots + f_t + f_{t+1} + \dots + f_{t+u}$, respectively, in the complexification $\mathfrak{g}_\mathbb{C}$, taking suitable conjugates of stabilizers in the complexification of G, we conclude that $Z_K(\frac{1}{2}(e+f))$ coincides with $Z_K(e)$. Finally, we may reduce the general case to the one just established by noting that $e_{t,u,\lambda}$ and $f_{t,u,\lambda}$ arise from $e_1+\dots+e_t-e_{t+1}-\dots-e_{t+u}$ and $f_1+\dots+f_t-f_{t+1}-\dots-f_{t+u}$ respectively as the result of the adjoint action with a suitable element from the constituent $A_{\mathfrak p}$ in the Iwasawa-decomposition $G=KA_{\mathfrak p}N$. Consequently $Z_K(\frac{1}{2}(e+f))$ coincides with $Z_K(e)$ in the general case.

We conclude that the projection from $Ke_{t,u,\lambda}$ onto its image in $\mathfrak p$ is a diffeomorphism. Furthermore, with the above normalization, the orbits $K(\frac{1}{2}(e_{t,u,\lambda}+f_{t,u,\lambda}))\subseteq\mathfrak p$ $(0<\lambda_1\leq\lambda_2\leq\dots\leq\lambda_t$ and $0<\lambda_{t+1}\leq\dots\leq\lambda_{t+u})$ are mutually disjoint, and their union coincides with the $K^{\mathbb C}$-orbit of $\frac{1}{2}(e_{t,u}+f_{t,u})$ in $\mathfrak p$ as may be see from writing out the $K^{\mathbb C}$-action on $\mathfrak p$ in terms of the polar decomposition $K^{\mathbb C}=K\exp(i\mathfrak k)$ of $K^{\mathbb C}$. In particular, given $0<\lambda_1\leq\lambda_2\leq\dots\leq\lambda_t$ and $0<\lambda_{t+1}\leq\dots\leq\lambda_{t+u}$, for some $X\in\mathfrak k$, we have

$$e_{t,u,\lambda}+f_{t,u,\lambda}=\exp(iX)(e_{t,u}+f_{t,u}).$$

Hence, for $0\leq t,u\leq r$ with $t+u\leq r$, the projection from $\mathcal O_{t,u}=Ge_{t,u}$ onto its image in $\mathfrak p$ is a diffeomorphism onto the $K^{\mathbb C}$-orbit of $\frac{1}{2}(e_{t,u}+f_{t,u})$ in $\mathfrak p$. In particular, $\mathcal O_{t,u}$ is pseudoholomorphic as asserted. □

REMARK. Let L_r be the (reductive) centralizer of h^r in the adjoint group G and let $K_r=K\cap L_r$. The union of the orbits $\mathcal O_{t,u}$ $(0\leq t,u\leq r,\ t+u\leq r)$ equals $K\mathfrak n_r^+\subseteq\mathfrak g$, and the projection to $\mathfrak p$ is induced by the canonical surjection from $K\times_{K_r}\mathfrak n_r^+$ to $\mathfrak p$ given by the assignment to $(x,\alpha)\in K\times\mathfrak n_r^+$ of $x\overline{\alpha}x^{-1}\in\mathfrak p$ where $\overline{\alpha}\in\mathfrak p$ refers to the image of $\alpha\in\mathfrak n_r^+$ under the projection to $\mathfrak p$. From this observation, one may as well deduce that the projection from $\mathcal O_{t,u}=Ge_{t,u}$ onto its image in $\mathfrak p$ is a diffeomorphism onto the $K^{\mathbb C}$-orbit of $\frac{1}{2}(e_{t,u}+f_{t,u})$ in $\mathfrak p$. For example, for $\mathfrak g=\mathfrak{sp}(\ell,\mathbb R)$, the real rank r equals ℓ, $\mathfrak n_r^+$ amounts to the space $\mathrm S^2[\mathbb R^\ell]$ of real symmetric $\ell\times\ell$-matrices, $\mathfrak p$ to that of complex symmetric $\ell\times\ell$-matrices $\mathrm S^2[\mathbb C^\ell]$, associating $\overline{\alpha}\in\mathfrak p$ to $\alpha\in\mathfrak n_r^+$ amounts to associating to a real symmetric $\ell\times\ell$-matrix the same matrix, viewed as a complex symmetric $\ell\times\ell$-matrix, and the canonical surjection from $K\times_{K_r}\mathfrak n_r^+$ to $\mathfrak p$ amounts to the familiar map from $\mathrm U(\ell)\times_{\mathrm O(\ell,\mathbb R)}\mathrm S^2[\mathbb R^\ell]$ to $\mathrm S^2[\mathbb C^\ell]$. The restriction of this map to the $\mathrm U(\ell)$-span of any $\mathrm O(\ell,\mathbb R)$-orbit of non-degenerate real symmetric $\ell\times\ell$-matrices is a diffeomorphism onto the space of non-degenerate complex symmetric $\ell\times\ell$-matrices.

PROOF OF THEOREM 3.3.3. Lemma 3.3.6 says that the orbits $\mathcal O_{t,u}$ are pseudoholomorphic. It remains to justify the claim as to the closures of the orbits $\mathcal O_{t,u}$ and to explain how the holomorphic and antiholomorphic orbits are singled out. To this end we observe that every nilpotent orbit of the semisimple Lie algebra of hermitian type $((\mathfrak{sl}(2,\mathbb R)^r,(z_1,\dots,z_r))$ is pseudoholomorphic; these orbits are generated by the $(\varepsilon_1e_1,\dots,\varepsilon_re_r)$'s where $\varepsilon_j=-1,0,+1$ $(1\leq j\leq r)$. The classification of holomorphic orbits of $\mathfrak{sl}(2,\mathbb R)$ reproduced in (3.2.2) above entails at once that the orbits involving only $\varepsilon_j=0,+1$ are the holomorphic ones and that, likewise, the orbits involving only $\varepsilon_j=-1,0$ are the antiholomorphic ones. Furthermore, when $\mathcal O$ is the $(\mathrm{SL}(2,\mathbb R)^r)$-orbit of $(\varepsilon_1e_1,\dots,\varepsilon_re_r)$, the orbits in the closure of $\mathcal O$ are precisely the orbits generated by those r-tuples which arise from $(\varepsilon_1e_1,\dots,\varepsilon_re_r)$ by replacing some $\varepsilon_j\neq0$ with zero. However, within $\mathfrak g$, the relative Weyl group $W_{\mathfrak p}\cong(\mathbb Z/2)^r\rtimes S_r$ permutes the r copies of $\mathfrak{sl}(2,\mathbb R)$ and hence in particular the entries of the $(\varepsilon_1e_1,\dots,\varepsilon_re_r)$ whence, within $\mathfrak g$, what remains invariant under the relative Weyl group, are the rank $t+u$ and the signature $t-u$. Thus an orbit $\mathcal O'$

distinct from $\mathcal{O}_{t,u}$ is in the closure of $\mathcal{O}_{t,u}$ if and only if $\mathcal{O}' = \mathcal{O}_{t',u'}$ for some $t' \leq t$ and $u' \leq u$ with $t'+u' < t+u$. Furthermore, since the complex structure J_z on $\mathfrak{p}$ is K-invariant, the orbits $\mathcal{O}_{t,0}$ $(t \leq r)$ and $\mathcal{O}_{0,u}$ $(u \leq r)$ are precisely the holomorphic and antiholomorphic ones, respectively.

Finally we establish the fact that the orbits $\mathcal{O}_{t,u}$ $(0 \leq t, u \leq r,\ t+u \leq r)$ exhaust the pseudoholomorphic nilpotent orbits. Inspection of the possible nilpotent orbits in $\mathfrak{g}$ shows that, given a nilpotent orbit which is different from one of the kind $\mathcal{O}_{t,u}$, the projection to the constituent $\mathfrak{p}$ of the Cartan decomposition of $\mathfrak{g}$ is no longer injective. For the classical cases we postpone the details to the proofs of Theorems 3.5.3 and 3.6.2 below. We now settle the two exceptional cases $\mathfrak{e}_{6(-14)}$ and $\mathfrak{e}_{7(-25)}$ by inspection. As before, G denotes the adjoint group of the corresponding simple Lie algebra $\mathfrak{g}$ under discussion.

The Lie algebra $\mathfrak{g} = \mathfrak{e}_{6(-14)}$ has 12 non-zero nilpotent orbits. The orbits numbered (1) to (5) in Table X of [**12**] (p. 512) are precisely the pseudoholomorphic ones; with the appropriate choice of H-element, the orbits (1) and (3) coincide with the orbits Ge^1 and Ge^2, respectively, the orbits (2) and (4) are antiholomorphic and amount to the orbits $G(-e^1)$ and $G(-e^2)$, while the orbit (5) is our pseudoholomorphic nilpotent orbit $G(e^1 - e^2)$ which is neither holomorphic nor antiholomorphic. There are no other pseudoholomorphic nilpotent orbits in $\mathfrak{e}_{6(-14)}$. Indeed, in this case, $\dim_{\mathbb{C}} \mathfrak{p} = 16$, and column 4) of Table X of [**12**] gives the complex dimensions of the stabilizers of the corresponding $K^{\mathbb{C}}$-orbits in $\mathfrak{p}$ where $K \cong \mathrm{Spin}_{\mathrm{c}}(10)$ is the maximal compact subgroup of the adjoint group G. Since $\dim K^{\mathbb{C}} = 46$, inspection of the dimensions of the stabilizers shows that, under the projection from $\mathfrak{g}$ to $\mathfrak{p}$, none of the orbits (6)–(12) can map diffeomorphically onto its image in $\mathfrak{p}$.

The Lie algebra $\mathfrak{g} = \mathfrak{e}_{7(-25)}$ has 22 non-zero nilpotent orbits. The orbits numbered (1) to (9) in Table XIII of [**12**] (p. 516) are precisely the pseudoholomorphic ones; with the appropriate choice of H-element, the orbits (1), (3), and (6) coincide with, respectively, the orbits Ge^1, Ge^2, and Ge^3, the orbits (2), (4), and (7) are antiholomorphic and amount to, respectively, the orbits $G(-e^1)$, $G(-e^2)$, and $G(-e^3)$, while the remaining orbits (5), (8), (9) amount to our pseudoholomorphic nilpotent orbits respectively $G(e^1 - e^2)$, $G(e^1 + e^2 - e^3)$, $G(e^1 - e^2 - e^3)$, which are neither holomorphic nor antiholomorphic. There are no other pseudoholomorphic nilpotent orbits in $\mathfrak{e}_{7(-25)}$. Indeed, in this case, $\dim_{\mathbb{C}} \mathfrak{p} = 27$, and the complex dimensions of the stabilizers of the corresponding $K^{\mathbb{C}}$-orbits in $\mathfrak{p}$ may be found in column 4) of Table X of [**12**] where K is the maximal compact subgroup of the adjoint group G; the group K is locally isomorphic to $E_{6(-78)} \times \mathrm{SO}(2, \mathbb{R})$ ($E_{6(-78)}$ being a compact form of E_6) and $\dim_{\mathbb{R}} K = 79$. Inspection of the dimensions of the stabilizers shows that, under the projection from $\mathfrak{g}$ to $\mathfrak{p}$, none of the orbits (10)–(22) can map diffeomorphically onto its image in $\mathfrak{p}$. □

PROOF OF THE ADDENDUM. Let z and $\widetilde{z}$ be two H-elements, and use the notation $\widetilde{e}_j$, $\widetilde{e}_{t,u}$, etc. with reference to $\widetilde{z}$. Then, for $0 \leq t, u \leq r$, $t + u \leq r$, $\mathcal{O}_{t,u} = Ge_{t,u} = G\widetilde{e}_{t,u} = \widetilde{\mathcal{O}_{t,u}}$ provided z and $\widetilde{z}$ are in the same component of the space of H-elements, and $\mathcal{O}_{t,u} = Ge_{t,u} = G\widetilde{e}_{u,t} = \widetilde{\mathcal{O}_{u,t}}$ otherwise. □

REMARK 3.3.7. In the standard cases, the holomorphicity of the orbits $\mathcal{O}_{t,0}$ $(1 \leq t \leq r)$, in particular, the assertion in the above proof of Lemma 3.3.6 involving the stabilizers, will also be a consequence of Theorem 5.3.3 below.

COROLLARY 3.3.8. *The projection from the closure $\overline{\mathcal{O}}$ of a pseudoholomorphic nilpotent orbit $\mathcal{O}$ to the constituent $\mathfrak{p}$ of the Cartan decomposition is injective if and only if $\mathcal{O}$ is holomorphic or antiholomorphic.*

PROOF. Indeed, a pseudoholomorphic nilpotent orbit $\mathcal{O}$ is one of the kind $\mathcal{O}_{t,u}$. If $u = 0$ or $t = 0$, for each $0 \leq s < \mathrm{rank}(\mathcal{O}_{t,u})$, there is only a single rank s nilpotent orbit in the closure $\overline{\mathcal{O}_{t,u}}$. If $u \neq 0$ and $t \neq 0$, the projection from the closure $\overline{\mathcal{O}_{t,u}}$ to $\mathfrak{p}$ is no longer injective since the closure contains at least two orbits of the same rank but having different signature; the projection to $\mathfrak{p}$ identifies the two. □

COROLLARY 3.3.9. *An orbit which belongs to the closure of a (pseudo-) holomorphic nilpotent orbit is necessarily (pseudo)holomorphic, and the closure of a (pseudo)holomorphic nilpotent orbit is a union of finitely many (pseudo-) holomorphic nilpotent orbits.*

PROOF. This is an almost immediate consequence of Theorem 3.3.3. □

PROOF OF THEOREM 3.2.1. This follows readily from Theorem 3.3.3. We leave the details to the reader. □

Let $(\mathfrak{g}, z)$ be a simple Lie algebra of hermitian type and real rank r; for $1 \leq t \leq r$, we will henceforth denote the holomorphic nilpotent orbit $\mathcal{O}_{t,0}$ by $\mathcal{O}_t$. Theorem 3.3.3 entails that the holomorphic nilpotent orbits $\mathcal{O}_0, \ldots, \mathcal{O}_r$ are linearly ordered in such a way that

$$\{0\} = \mathcal{O}_0 \subseteq \overline{\mathcal{O}_1} \subseteq \ldots \subseteq \overline{\mathcal{O}_r}. \tag{3.3.10}$$

Borrowing and extending terminology from [**53**], where the regular (complex) nilpotent orbit is said to be principal, we will refer to the top orbit $\mathcal{O}_r$ as the *principal holomorphic* nilpotent orbit. The principal holomorphic nilpotent (in our sense) may be viewed as the *regular holomorphic nilpotent orbit*; it is in particular unique, and its closure contains every holomorphic nilpotent orbit.

THEOREM 3.3.11. *For the principal holomorphic nilpotent orbit $\mathcal{O}_r$, the composite of the projection from the closure $\overline{\mathcal{O}_r}$ to $\mathfrak{p}$ with the complex linear isomorphism from $\mathfrak{p}$ to $\mathfrak{p}^+$ is a homeomorphism $\overline{\mathcal{O}_r} \to \mathfrak{p}^+$. Under this homeomorphism, the G-orbit stratification of $\overline{\mathcal{O}_r}$ passes to the $K^{\mathbb{C}}$-orbit stratification of $\mathfrak{p}^+$. Thus, for $1 \leq s \leq r$, restricted to $\mathcal{O}_s = Ge^s$, this homeomorphism is a K-equivariant diffeomorphism from $\mathcal{O}_s$ onto the $K^{\mathbb{C}}$-orbit in $\mathfrak{p}^+$ of $x_s = \frac{1}{2}(e^s + f^s - ih^s) \in \mathfrak{p}^+$.*

PROOF. This is a consequence of Lemma 3.3.6, except perhaps the fact that the indicated map is onto $\mathfrak{p}^+$; the latter, in turn, is implied by the observation that $\mathfrak{p}^+$ is the disjoint union of its $K^{\mathbb{C}}$-orbits. □

REMARK 3.3.12. In [**25**] (p. 185) a nilpotent orbit $\mathcal{O}$ of $\mathfrak{g}$ is said to be of convex type provided it is contained in a proper generating invariant cone, and a complete classification of orbits of convex type is given. The orbits of convex type in this sense are precisely the holomorphic and antiholomorphic ones.

REMARK 3.3.13. It is known (Theorem 1.9 in [**83**]) that, for a general real semisimple Lie algebra $\mathfrak{g}$, the assignment to (e, f, h) of $\frac{1}{2}(e + f - ih)$ where (e, f, h) runs through invariant $\mathfrak{sl}(2)$-triples in $\mathfrak{g}$ yields a bijective correspondence $\mathcal{O} \mapsto c(\mathcal{O})$, referred to in the literature as *Kostant-Sekiguchi correspondence*, between nilpotent (G-)orbits in $\mathfrak{g}$ and nilpotent $K_{\mathbb{C}}$-orbits in $\mathfrak{p}_{\mathbb{C}}$ (i. e. $K_{\mathbb{C}}$-orbits in $\mathfrak{p}_{\mathbb{C}}$ which arise as intersection of a nilpotent orbit in $\mathfrak{g}_{\mathbb{C}}$ with $\mathfrak{p}_{\mathbb{C}}$). In particular, for a semisimple Lie algebra $(\mathfrak{g}, z)$ of hermitian type, a little thought reveals that the holomorphic nilpotent orbits $\mathcal{O}$ in $\mathfrak{g}$ are precisely those which have the property that the orbit $c(\mathcal{O})$ lies in $\mathfrak{p}^+$. In [**89**], a nilpotent orbit $\mathcal{O}$ of $\mathfrak{g}$ is defined to be holomorphic provided $c(\mathcal{O})$ lies in $\mathfrak{p}^+$; it is then observed that, for a nilpotent orbit $\mathcal{O}$ having the property that $c(\mathcal{O})$ lies in $\mathfrak{p}^+$, the projection from $\mathcal{O}$ to $\mathfrak{p}$ is a diffeomorphism onto its image. The argument in [**89**] relies on the fact, established in [**89**] that, for an arbitrary nilpotent orbit $\mathcal{O}$, the Kostant-Sekiguchi correspondence may be realized by a diffeomorphism from $\mathcal{O}$ to $c(\mathcal{O})$; this diffeomorphism involves a certain flow constructed by Kronheimer in [**57**] and is referred to in the literature as *Kronheimer-Vergne diffeomorphism*. Our approach avoids the detour via this diffeomorphism.

REMARK 3.3.14. The classification of nilpotent orbits in the real exceptional Lie algebras given in [**12**] (coming into play in the proof of Theorem 3.3.3) relies on the Kostant-Sekiguchi correspondence. We now indicate briefly how, in the two exceptional cases, the holomorphic nilpotent orbits can be classified without reference to the Kostant-Sekiguchi correspondence: Over the complex numbers, the classification of nilpotent orbits is in terms of "regular" subalgebras of $\mathfrak{g}$ of minimal rank containing a representative of the nilpotent orbit being classified, the requisite subalgebra being written as L_1 in [**13**] and $L(x)$ in [**67**]. These subalgebras actually play a role similar to that of the indecomposable types in [**7**] for the classical cases which we exploited for the proofs of Theorems 3.5.3 and 3.6.2. As before, G refers to the adjoinnt group of the Lie algebra $\mathfrak{g}$ under discussion.
$\mathfrak{g} = \mathfrak{e}_{6(-14)}$: The orbits Ge^1 and Ge^2 complexify to the orbits in $\mathfrak{g}_{\mathbb{C}}$ given in terms of the subalgebras L_1 and/or $L(x)$ written as A_1 and $2A_1$, respectively, on p. 152 of [**13**] and in Table 1 on p. 447 of [**67**]. Inspection of these tables shows that no other nilpotent orbit in $\mathfrak{g}$ can be holomorphic: given any such orbit, it complexifies to an orbit involving a regular subalgebra of minimal rank which does *not* arise by complexification from a reductive Lie algebra of hermitian type. Hence $\mathfrak{g} = \mathfrak{e}_{6(-14)}$ has no holomorphic nilpotent orbits other than $\{0\}, \mathcal{O}_1 = Ge^1, \mathcal{O}_2 = Ge^2$. A slight extension of this reasoning shows that the orbits $\mathcal{O}_{t,u}$ where $0 \leq t, u \leq 2$ and $t + u \leq 2$ constitute a complete list of pseudoholomorphic nilpotent orbits.
$\mathfrak{g} = \mathfrak{e}_{7(-25)}$: The orbits Ge^1, Ge^2, and Ge^3 complexify to the orbits in $\mathfrak{g}_{\mathbb{C}}$ given in terms of the subalgebras L_1 and/or $L(x)$ written, respectively, as A_1, $2A_1$, $(3A_1)''$, on p. 153 of [**13**] and in Table 2 on p. 447 of [**67**]. Inspection of these tables shows that no other nilpotent orbit in $\mathfrak{g}$ can be holomorphic except possibly one which complexifies to an orbit in $\mathfrak{g}_{\mathbb{C}}$ given in terms of the subalgebra L_1 and/or $L(x)$ written as $(3A_1)'$. However, inspection of Table 9 on p. 454 of [**67**] reveals that the stabilizer of this orbit has dimension 69 whence the dimension of this orbit is strictly larger than that of $\mathfrak{p}$ and hence this orbit cannot possibly arise from a (real) pseudoholomorphic nilpotent orbit. Actually, this orbit does not arise from any real orbit at all. Hence $\mathfrak{g} = \mathfrak{e}_{7(-25)}$ has no other holomorphic nilpotent orbits than $\{0\}, \mathcal{O}_1 = Ge^1, \mathcal{O}_2 = Ge^2, \mathcal{O}_3 = Ge^3$. A slight extension of this reasoning shows that the orbits $\mathcal{O}_{t,u}$ where $0 \leq t, u \leq 3$ and $t + u \leq 3$ constitute a complete list of pseudoholomorphic nilpotent orbits.

3.4. NON-NILPOTENT PSEUDOHOLOMORPHIC ORBITS. Let $(\mathfrak{g}, z)$ be a simple Lie algebra of hermitian type and split rank r, with Cartan decomposition $\mathfrak{g} = \mathfrak{k} \oplus \mathfrak{p}$.

Let $\varepsilon > 0$ and let $\mathcal{O}^+(\varepsilon)$ and $\mathcal{O}^-(\varepsilon)$ be the (semisimple elliptic) orbits of $2\varepsilon z$ and $-2\varepsilon z$, respectively; these generalize the non-nilpotent pseudoholomorphic orbits in $\mathfrak{sl}(2,\mathbb{R})$, cf. (3.2.2) above.

THEOREM 3.4.1. *An orbit of the kind $\mathcal{O}^+(\varepsilon)$ is holomorphic and one of the kind $\mathcal{O}^-(\varepsilon)$ is antiholomorphic.*

PROOF. Indeed, for simplicity, let $\varepsilon = \frac{1}{2}$, so that $\mathcal{O}^+(\varepsilon) = Gz$. Since $G = KA_{\mathfrak{p}}K$, the orbit $\mathcal{O}^+(\varepsilon)$ may be written as $\mathcal{O}^+(\varepsilon) = Gz = KA_{\mathfrak{p}}z$. We now assert that the projection from Gz to $\mathfrak{p}$ is a diffeomorphism. In order to see this, exploiting Proposition 3.3.2, pick a maximal abelian subalgebra $\mathfrak{a}_{\mathfrak{p}}$ of $\mathfrak{p}$, let (e_1, f_1, h_1), $\dots$,(e_r, f_r, h_r) be the corresponding system of H_1-triples so that, in particular, the elements $h_1, \dots, h_r$ constitute a basis of $\mathfrak{a}_{\mathfrak{p}}$, and consider the resulting H_1-embedding of $(\mathfrak{sl}(2,\mathbb{R})^r, Z^r)$ into $(\mathfrak{g}, z)$. Let $\widetilde{z} = \frac{1}{2}(e_1 - f_1 + \dots + e_r - f_r)$. Notice that, in view of Proposition 3.3.2(3), in the regular case, $\widetilde{z} = z \in \mathfrak{g}$. Whether or not $(\mathfrak{g}, z)$ is regular, since the embedding of $(\mathfrak{sl}(2,\mathbb{R})^r, Z^r)$ into $(\mathfrak{g}, z)$ is an H_1-embedding, we have $[z, h_j] = [\widetilde{z}, h_j]$ for every j $(1 \leq j \leq r)$. Consequently $A_{\mathfrak{p}}$ leaves the point $z - \widetilde{z}$ of $\mathfrak{g}$ invariant. Hence $A_{\mathfrak{p}}z = A_{\mathfrak{p}}\widetilde{z} + z - \widetilde{z}$. Furthermore, since $\mathfrak{a}_p$ is the span of $h_1, \dots, h_r$, the orbit $A_{\mathfrak{p}}\widetilde{z}$ consists of the elements

$$\frac{1}{2}(c_1 + \dots + c_r) \in \mathfrak{g}$$

where $c_j = \mathrm{ch}(\lambda_j)(e_j - f_j) + \mathrm{sh}(\lambda_j)(e_j + f_j)$, for $1 \leq j \leq r$, with $\lambda_1, \dots, \lambda_r \in \mathbb{R}$. Thus the orbit $A_{\mathfrak{p}}z$ consists of the points

$$\frac{1}{2}(c_1 + \dots + c_r) + z - \widetilde{z} \in \mathfrak{g}.$$

Since $e_j - f_j \in \mathfrak{k}$ $(1 \leq j \leq r)$ and since $z - \widetilde{z} \in \mathfrak{k}$, the projection from the orbit $A_{\mathfrak{p}}z$ to $\mathfrak{p}$ sends each $\frac{1}{2}(c_1 + \dots + c_r) + z - \widetilde{z}$ to

$$\frac{1}{2}(\mathrm{sh}(\lambda_1)(e_1 + f_1) + \dots + \mathrm{sh}(\lambda_r)(e_r + f_r)),$$

and this projection is a diffeomorphism onto its image. Since

$$Gz = KA_{\mathfrak{p}}z = K(A_{\mathfrak{p}}\widetilde{z} + z - \widetilde{z}) = K(A_{\mathfrak{p}}\widetilde{z} - \widetilde{z}) + z,$$

the elements $\frac{1}{2}(c_1 + \dots + c_r) + z - \widetilde{z}$ constitute a complete set of representatives of the K-orbits of Gz. Furthermore, it is well known that $\mathfrak{p} = \cup_{y \in K}\mathrm{Ad}(y)\mathfrak{a}_p$ whence the elements

$$\frac{1}{2}(\mathrm{sh}(\lambda_1)(h_1) + \dots + \mathrm{sh}(\lambda_r)(h_r))$$

constitute a complete set of representatives of the K-orbits of $\mathfrak{p}$; this fact entails that the elements

$$\frac{1}{2}(\mathrm{sh}(\lambda_1)(e_1 + f_1) + \dots + \mathrm{sh}(\lambda_r)(e_r + f_r))$$

constitute a complete set of representatives of the K-orbits of $\mathfrak{p}$ as well since, for $1 \leq j \leq r$, $J_z(e_j + f_j) = h_j$, and since the operator J_z may be realized by the adjoint action with an element of K. Consequently the projection from the orbit Gz to $\mathfrak{p}$ is a bijection onto its image, in fact, a diffeomorphism onto $\mathfrak{p}$. Moreover, since the orbit $\mathcal{O}^+(\varepsilon)(1) \times \dots \times \mathcal{O}^+(\varepsilon)(r)$ in $\mathfrak{sl}(2,\mathbb{R})^r$ is holomorphic where $\mathcal{O}^+(\varepsilon)(j)$ refers to the corresponding orbit in the j'th copy of $\mathfrak{sl}(2,\mathbb{R})$ in $\mathfrak{sl}(2,\mathbb{R})^r$, cf. Proposition

3.3.2, we conclude that the orbit Gz is holomorphic in $\mathfrak{g}$. Likewise, the orbit $G(-z)$ is antiholomorphic in $\mathfrak{g}$. □

THEOREM 3.4.2. *The only non-nilpotent pseudoholomorphic orbits are the semisimple holomorphic and antiholomorphic ones, i. e. those of the kind* $\mathcal{O}^+(\varepsilon)$ *and* $\mathcal{O}^-(\varepsilon)$.

Thus there are no other holomorphic orbits than the semisimple and the nilpotent ones, and the latter ones are classified in Theorem 3.3.3 above.

PROOF. Given a (co)adjoint orbit $\mathcal{O}$ of a general real or complex semisimple Lie algebra $\mathfrak{g}$, the closure $\overline{\mathcal{O}}$ contains a single semisimple orbit $\mathcal{O}_{\mathrm{ss}}$, and any other orbit in $\overline{\mathcal{O}}$ fibers over $\mathcal{O}_{\mathrm{ss}}$ with typical fiber a nilpotent orbit for some reductive subalgebra $\mathfrak{g}'$ of $\mathfrak{g}$ which is a proper subalgebra when $\mathcal{O}_{\mathrm{ss}}$ is non-trivial. More precisely, the assignment to $x \in \overline{\mathcal{O}}$ of the semisimple constituent s in its Jordan decomposition $x = s+u$ is a fiber bundle map $\pi\colon \overline{\mathcal{O}} \to \mathcal{O}_{\mathrm{ss}}$; for $s \in \mathcal{O}_{\mathrm{ss}}$, the stabilizer $Z_G(s)$ of s in the adjoint group G is reductive, and the fiber $\pi^{-1}(s)$ is (isomorphic to) the closure $\overline{(Z_G(s))u} \subseteq \mathfrak{z}_{\mathfrak{g}}(s)$ of the orbit $(Z_G(s))u$ in $\mathfrak{z}_{\mathfrak{g}}(s) = \mathrm{Lie}(Z_G(s))$ of any u with $x = s + u \in \mathcal{O}$. Thus, given a (real) simple Lie algebra of hermitian type $(\mathfrak{g}, z)$, let $\mathcal{O}$ be a non-nilpotent pseudoholomorphic orbit, and let $\mathcal{O}_{\mathrm{ss}}$ be the unique semisimple orbit in $\overline{\mathcal{O}}$. Since $\mathfrak{k}_{\mathbb{C}}$ has the same rank as $\mathfrak{g}_{\mathbb{C}}$ and since the span of $e_1 - f_1, \ldots, e_r - f_r$ is a Cartan subalgebra of $\mathfrak{k}_{\mathbb{C}}$ and hence of $\mathfrak{g}_{\mathbb{C}}$, there are complex numbers $\lambda_1, \ldots, \lambda_s$ $(s \leq r)$ such that the complexification $\mathcal{O}^{\mathbb{C}}_{\mathrm{ss}}$ is the $G_{\mathbb{C}}$-orbit of $w = \lambda_1(e_1 - f_1) + \cdots + \lambda_s(e_s - f_s)$. Since $\mathcal{O}$ is pseudoholomorphic, the projection from $\mathcal{O}$ to $\mathfrak{p}$ is a diffeomorphism onto its image, and the induced tangent map $\mathrm{T}_{w+u}(\mathcal{O}^{\mathbb{C}}) \to \mathfrak{p}_{\mathbb{C}}$ is injective for any u with $w + u \in \mathcal{O}^{\mathbb{C}}$, the complexification of $\mathcal{O}$. Moreover, the tangent space $\mathrm{T}_{w+u}(\mathcal{O}^{\mathbb{C}})$ contains the tangent space $\mathrm{T}_w(\mathcal{O}^{\mathbb{C}}_{\mathrm{ss}})$. Hence $[y, w] = 0$ for every $y \in \mathfrak{k}_{\mathbb{C}}$. But $w \in \mathfrak{k}_{\mathbb{C}}$ whence w lies in the center of $\mathfrak{k}_{\mathbb{C}}$, that is, $w = \lambda z$ for some non-zero complex number λ. Consequently $\mathcal{O}_{\mathrm{ss}}$ is the G-orbit of εz for some non-zero real number ε, that is to say, $\mathcal{O}_{\mathrm{ss}}$ is a semisimple holomorphic or antiholomorphic orbit, and for dimensional reasons, $\mathcal{O}_{\mathrm{ss}}$ coincides with $\mathcal{O}$. □

ADDENDUM. *For a semisimple orbit of a simple Lie algebra* $(\mathfrak{g}, z)$ *of hermitian type, the property of being holomorphic or antiholomorphic does not depend on the choice of* H*-element* z.

This is established in much the same way as the Addendum to Theorem 3.3.3.

REMARK 3.4.3. The holomorphic discrete series representations arise from holomorphic quantization on integral holomorphic semisimple orbits (but with reference to a complex structure different from that coming from the projection to $\mathfrak{p}$). Thus our terminology "holomorphic semisimple orbit" is consistent with established terminology in the literature.

3.5. THE STANDARD CASES $\mathfrak{sp}(n, \mathbb{R})$, $\mathfrak{su}(p, q)$, $\mathfrak{so}^*(2n)$.

3.5.1. EXPLICIT DESCRIPTIONS OF THESE LIE ALGEBRAS. We shall need them for ease of exposition and later reference. Thus, let $\mathbb{K} = \mathbb{R}, \mathbb{C}, \mathbb{H}$, and let V be an n-dimensional *right* $\mathbb{K}$-vector space, endowed with a positive definite hermitian form $(\cdot, \cdot)$, that is

$$(x, y) = \overline{(y, x)}, \quad (x, y\lambda) = (x, y)\lambda, \quad (x\lambda, y) = \overline{\lambda}(x, y), \quad x, y \in V,\ \lambda \in \mathbb{K},$$

where the bar indicates conjugation as usual. For right vector spaces over the quaternions, requiring linearity in the second variable is more appropriate than the more usual requirement of linearity in the first variable. After a suitable choice of basis, V gets identified with $\mathbb{K}^n$, the algebra $\mathrm{End}_{\mathbb{K}}(V)$ may be identified with the (left $\mathbb{K}$-vector) space of $(n \times n)$-matrices with entries from $\mathbb{K}$ which act on column vectors from the left in the usual fashion, and $(\cdot,\cdot)$ is given by $(x,y) = \sum \overline{x_j} y_j$ $(x, y \in V)$; in particular, composition of endomorphisms now corresponds to multiplication of matrices. The Lie group $\mathrm{U}(V,(\cdot,\cdot))$ of $(\cdot,\cdot)$-isometries and its Lie algebra $\mathfrak{u}(V,(\cdot,\cdot))$ amount to $\mathrm{O}(n,\mathbb{R})$ and $\mathfrak{so}(n,\mathbb{R})$ for $\mathbb{K}=\mathbb{R}$, to $\mathrm{U}(n)$ and $\mathfrak{u}(n)$ for $\mathbb{K}=\mathbb{C}$, and to $\mathrm{Sp}(n)$ and $\mathfrak{sp}(n)$ for $\mathbb{K}=\mathbb{H}$.

The requisite additional ingredient is a ($\mathbb{K}$-linear) complex structure J_V on V which is compatible with $(\cdot,\cdot)$, that is, $J_V \in \mathfrak{u}(V,(\cdot,\cdot))$ or, equivalently, for every $u, v \in V$, $(J_V u, v) + (u, J_V v) = 0$. Define the non-degenerate skew-hermitian form $\mathcal{B}$ on V by $\mathcal{B}(u,v) = (J_V u, v)$, where $u, v \in V$. We write $\mathrm{U}(V,\mathcal{B})$ for the Lie group of $\mathcal{B}$-isometries, $\mathrm{U}(V,\mathcal{B})^0$ for its connected component of the identity, and $\mathfrak{u}(V,\mathcal{B}) = \mathrm{Lie}(\mathrm{U}(V,\mathcal{B}))$ for its Lie algebra (which consists of all $X \in \mathrm{End}_{\mathbb{K}}(V)$ satisfying $\mathcal{B}(Xv,w) + \mathcal{B}(v,Xw) = 0$ for $v, w \in V$). Thus $X \in \mathrm{End}_{\mathbb{K}}(V)$ lies in $\mathfrak{u}(V,\mathcal{B})$ if and only if the assignment

$$\mathcal{B}_X(v,w) = -\mathcal{B}(Xv,w) \quad (= -(J_V X v, w)), \quad v, w \in V, \tag{3.5.1.1}$$

yields a hermitian form $\mathcal{B}_X$ on V. Then J_V lies in $\mathfrak{u}(V,\mathcal{B})$, and $\mathcal{B}_{J_V} = (\cdot,\cdot)$ (the latter identity justifies, in particular, the minus sign in (3.5.1.1)). This sign also entails that, cf. what is said after Theorem 3.5.4, $\mathcal{B}_X$ being non-negative (in a sense to be explained) is equivalent to the holomorphicity of the orbit generated by X. The maximal compact subgroup of $\mathrm{U}(V,\mathcal{B})$ and, accordingly, the Lie algebra thereof, are the intersections $\mathrm{U}(V,\mathcal{B}) \cap \mathrm{U}(V,(\cdot,\cdot))$ and $\mathfrak{u}(V,\mathcal{B}) \cap \mathfrak{u}(V,(\cdot,\cdot))$, respectively. Further, with the standard notation X^* for the adjoint of $X \in \mathrm{End}_{\mathbb{K}}(V)$ so that $(Xu,v) = (u, X^* v)$ for every $u, v \in V$, we have $J_V^* = -J_V$, and $X \in \mathrm{End}_{\mathbb{K}}(V)$ lies in $\mathfrak{u}(V,\mathcal{B})$ if and only if $J_V X + X J_V = 0 \in \mathrm{End}_{\mathbb{K}}(V)$. Now $z_{\mathcal{B}} = \frac{1}{2} J_V$ is an H-element, so that $(\mathfrak{g}, z_{\mathcal{B}}) = (\mathfrak{u}(V,\mathcal{B}), z_{\mathcal{B}})$ is a simple Lie algebra of hermitian type in case $\mathbb{K}=\mathbb{R}$ or $\mathbb{K}=\mathbb{H}$, and a reductive Lie algebra of hermitian type in case $\mathbb{K}=\mathbb{C}$ whose semisimple part is simple and whose center is a copy of $\mathbb{R}$. We will write $\mathfrak{su}(V,\mathcal{B})$ for the corresponding simple Lie algebra (for $\mathbb{K}=\mathbb{R}$ and $\mathbb{K}=\mathbb{H}$ there is no difference between $\mathfrak{u}(V,\mathcal{B})$ and $\mathfrak{su}(V,\mathcal{B})$). With the notation r for the corresponding split rank and $1, \mathcal{I}, \mathcal{J}, \mathcal{K}$ for the unit quaternions, i. e. standard real basis of $\mathbb{H}$, we now recall the standard descriptions of the three classes of *standard* simple Lie algebras $(\mathfrak{g}, z_{\mathcal{B}})$ of hermitian type, where V is the *standard representation*, and where the Cartan decomposition is written as $\mathfrak{k} \oplus \mathfrak{p}$. Here and henceforth we will often write a matrix $\begin{bmatrix} \mathbf{v}_1 \\ \dots \\ \mathbf{v}_t \end{bmatrix}$ having row vectors $\mathbf{v}_1, \dots, \mathbf{v}_t$ in the form $[\mathbf{v}_1, \dots, \mathbf{v}_t]^\tau$ where the superscript τ stands for "transpose".

(3.5.1.2) $\mathbb{K}=\mathbb{R}$, $n = 2\ell$, $V = \mathbb{R}^n$, $r = \ell$, $J_V = \begin{bmatrix} 0 & -I_\ell \\ I_\ell & 0 \end{bmatrix}$,

$$\mathcal{B}(\mathbf{x},\mathbf{y}) = \sum_{j=1}^{\ell} (x_j y_{j+\ell} - x_{j+\ell} y_j), \ \mathbf{x}, \mathbf{y} \in V$$

$$\mathrm{U}(V,\mathcal{B}) = \mathrm{U}(V,\mathcal{B})^0 = \mathrm{Sp}(\ell,\mathbb{R})$$

$$\mathfrak{g} = \mathfrak{u}(V, \mathcal{B}) = \mathfrak{sp}(\ell, \mathbb{R})$$
$$= \left\{ \begin{bmatrix} A & B \\ C & -A^{\mathrm{t}} \end{bmatrix}; B^{\mathrm{t}} = B,\ C^{\mathrm{t}} = C,\ A, B, C \in \mathrm{M}_{\ell,\ell}(\mathbb{R}) \right\}$$
$$\mathfrak{k} = \mathfrak{u}(\ell) = \left\{ \begin{bmatrix} A & B \\ -B & A \end{bmatrix}; A^{\mathrm{t}} = -A,\ B^{\mathrm{t}} = B \right\}$$
$$\mathfrak{p} = \left\{ \begin{bmatrix} A & B \\ B & -A \end{bmatrix}; A^{\mathrm{t}} = A,\ B^{\mathrm{t}} = B \right\} = \mathrm{S}^2_{\mathbb{R}}[\mathbb{R}^\ell] \oplus \mathrm{S}^2_{\mathbb{R}}[\mathbb{R}^\ell]$$
$$\mathfrak{g} = \mathfrak{k} \oplus \mathfrak{p}: \quad \begin{bmatrix} A & B \\ C & -A^{\mathrm{t}} \end{bmatrix} = \begin{bmatrix} \frac{A-A^{\mathrm{t}}}{2} & \frac{B-C}{2} \\ \frac{C-B}{2} & \frac{A-A^{\mathrm{t}}}{2} \end{bmatrix} + \begin{bmatrix} \frac{A+A^{\mathrm{t}}}{2} & \frac{B+C}{2} \\ \frac{C+B}{2} & -\frac{A+A^{\mathrm{t}}}{2} \end{bmatrix}.$$

REMARK. Our present sign convention is guided by the standard description of angular momentum, that is, with the notation $\mathbf{x} = [\mathbf{q}_1, \mathbf{p}_1]^\tau$ and $\mathbf{y} = [\mathbf{q}_2, \mathbf{p}_2]^\tau$ where $\mathbf{q}_1, \mathbf{q}_2, \mathbf{p}_1, \mathbf{p}_2 \in \mathbb{R}^\ell$, the given expression for $\mathcal{B}(\mathbf{x}, \mathbf{y})$ equals the expression for $\omega([\mathbf{q}_1, \mathbf{p}_1]^\tau, [\mathbf{q}_2, \mathbf{p}_2]^\tau)$ for the standard symplectic structure ω on $\mathbb{R}^{2\ell}$ evaluated at these vectors, that is, the skew form $\mathcal{B}$ comes down to the standard symplectic form ω on $\mathbb{R}^{2\ell}$; for example, when $\ell = 1$, $\mathcal{B}([q^1, p^1]^\tau, [q^2, p^2]^\tau) = q^1p^2 - q^2p^1$. See also what is said in Section 5 below. When (V, J_V) is viewed as a flat Kähler manifold, the present sign for J_V etc. is consistent with the standard convention in Kähler geometry, to the effect that, in particular, when V is the ordinary plane, with coordinates x, y, the standard complex structure is given by $J\frac{\partial}{\partial x} = \frac{\partial}{\partial y}$ and $J\frac{\partial}{\partial y} = -\frac{\partial}{\partial x}$. The sign convention for J_V and $\mathcal{B}$ in (3.2.2) above with $V = \mathbb{R}^2$ and $J_V = E - F = \begin{bmatrix} 0 & 1 \\ -1 & 0 \end{bmatrix}$, as well as in Satake's book **[79]** and elsewhere in the literature, is opposite to the present one. Changing the sign convention amounts to interchanging holomorphic and antiholomorphic orbits. Under the circumstances of (3.2.2), the present sign convention may be realized via the $\mathfrak{sl}(2)$-triple $(-E, -F, H)$ in $\mathfrak{sl}(2, \mathbb{R})$, so that $\frac{1}{2}(F - E)$ is the corresponding H-element.

(3.5.1.3) $\mathbb{K} = \mathbb{C}$, $V = \mathbb{C}^n$, $n = p + q$, $p \geq q = r$, $J_V = i\begin{bmatrix} I_p & 0 \\ 0 & -I_q \end{bmatrix}$,

$$\mathcal{B}(\mathbf{x}, \mathbf{y}) = i\left(\sum_{j=p+1}^{p+q} x_j\overline{y}_j - \sum_{j=1}^{p} x_j\overline{y}_j\right), \ \mathbf{x}, \mathbf{y} \in V$$
$$\mathrm{U}(V, \mathcal{B}) = \mathrm{U}(V, \mathcal{B})^0 = \mathrm{U}(p, q)$$
$$\mathfrak{g} = \mathfrak{su}(V, \mathcal{B}) = \mathfrak{su}(p, q)$$
$$= \left\{ \begin{bmatrix} A & B \\ B^* & D \end{bmatrix}; A^* = -A,\ D^* = -D,\ \mathrm{tr}(A) + \mathrm{tr}(D) = 0 \right\}$$

where $A \in \mathrm{M}_{p,p}(\mathbb{C}), D \in \mathrm{M}_{q,q}(\mathbb{C}), B \in \mathrm{M}_{q,p}(\mathbb{C})$

$$\mathfrak{k} = (\mathfrak{u}(p) \oplus \mathfrak{u}(q)) \cap \mathfrak{su}(p, q),\ \mathfrak{p} = \left\{ \begin{bmatrix} 0 & B \\ B^* & 0 \end{bmatrix} \right\} \cong \mathrm{M}_{q,p}(\mathbb{C}).$$

When $p > q$, J_V lies in $\mathfrak{u}(p, q)$ but not in $\mathfrak{su}(p, q)$. However, with reference to the decomposition $\mathfrak{u}(p, q) = \mathfrak{z} \oplus \mathfrak{su}(p, q)$ where $\mathfrak{z} \cong \mathbb{R}$ is the center of $\mathfrak{u}(p, q)$, the element J_V decomposes as

$$J_V = i\frac{p - q}{p + q}I_{p+q} + J'_V, \quad \text{where} \quad J'_V = i\begin{bmatrix} \frac{2q}{p+q}I_p & 0 \\ 0 & -\frac{2p}{p+q}I_q \end{bmatrix},$$

and $\frac{1}{2}J'_V \in \mathfrak{su}(p,q)$ is an H-element; since, in case $p > q$, the constituent $i\frac{p-q}{p+q}I_{p+q}$ in the decomposition of J_V lies in the center of $\mathfrak{u}(p,q)$, it acts trivially on $\mathfrak{p}$ and the complex structure on $\mathfrak{p}$ given by $\frac{1}{2}\mathrm{ad}_{J'_V}$ within $\mathfrak{su}(p,q)$ is the same as that given by $\frac{1}{2}\mathrm{ad}_{J_V}$ within $\mathfrak{u}(p,q)$. See also Ex. III.2.1 on p. 98 of [**79**].
(3.5.1.4) $\mathbb{K} = \mathbb{H}$, $V = \mathbb{H}^n$, $r = [\frac{n}{2}]$, $J_V = \mathcal{J} I_n$

$$
\begin{aligned}
\mathcal{B}(\mathbf{x},\mathbf{y}) &= \sum_{j=1}^n \overline{x_j}\mathcal{J} y_j,\ \mathbf{x},\mathbf{y} \in V\\
\mathrm{U}(V,\mathcal{B}) &= \mathrm{O}^*(2n) \quad (\text{the non-compact dual of } \mathrm{O}(2n,\mathbb{R}))\\
\mathrm{U}(V,\mathcal{B})^0 &= \mathrm{SO}^*(2n) \quad (= \mathrm{SU}^-(n,\mathbb{H}))\\
\mathfrak{g} &= \mathfrak{u}(V,\mathcal{B}) = \mathfrak{so}^*(2n)\\
&= \left\{\begin{bmatrix} A & -B \\ \overline{B} & \overline{A}\end{bmatrix}; A^{\mathrm{t}} = -A,\ B^* = B,\ A, B \in \mathrm{M}_{n,n}(\mathbb{C})\right\}\\
\mathfrak{k} = \mathfrak{u}(n) &= \left\{\begin{bmatrix} A & -B \\ B & A\end{bmatrix} \in \mathfrak{so}^*(2n); A = \overline{A} = -A^{\mathrm{t}},\ B = \overline{B} = B^{\mathrm{t}}\right\}\\
\mathfrak{p} &= \left\{ i\begin{bmatrix} V & W \\ W & -V\end{bmatrix}; V^{\mathrm{t}} = -V,\ W^{\mathrm{t}} = -W,\ V, W \in \mathrm{M}_{n,n}(\mathbb{R})\right\}\\
&= i\Lambda^2_{\mathbb{R}}[\mathbb{R}^n] \oplus i\Lambda^2_{\mathbb{R}}[\mathbb{R}^n]\\
\mathfrak{g} = \mathfrak{k} \oplus \mathfrak{p} : &\quad \begin{bmatrix} A & -B \\ \overline{B} & \overline{A}\end{bmatrix} = \begin{bmatrix} \frac{A+\overline{A}}{2} & -\frac{B+\overline{B}}{2} \\ \frac{B+\overline{B}}{2} & \frac{A+\overline{A}}{2}\end{bmatrix} + \begin{bmatrix} \frac{A-\overline{A}}{2} & -\frac{\overline{B}-B}{2} \\ \frac{\overline{B}-B}{2} & \frac{\overline{A}-A}{2}\end{bmatrix}.
\end{aligned}
$$

REMARKS. (i) In each of these cases, the complex structure $J_{z_{\mathcal{B}}}$ on $\mathfrak{p}$ is simply given by matrix multiplication by J_V from the left.
(ii) In case (3.5.1.3), the form $\mathcal{B}$, written out as $\mathcal{B} = \mathcal{B}_1 + i\mathcal{B}_2$ with real forms $\mathcal{B}_1$, $\mathcal{B}_2$, has $\mathfrak{u}(V,\mathcal{B}_1) = \mathfrak{sp}(p+q,\mathbb{R})$ and $\mathfrak{u}(V,\mathcal{B}_2) = \mathfrak{so}(2p,2q)$ whence $\mathfrak{u}(p,q) = \mathfrak{so}(2p,2q) \cap \mathfrak{sp}(p+q,\mathbb{R})$. The resulting inclusion

$$\mathfrak{su}(p,q) \to \mathfrak{sp}(p+q,\mathbb{R}) \tag{3.5.1.5}$$

is an H_1-embedding. Alternatively, $\mathfrak{u}(p,q)$ arises as the centralizer of

$$\begin{bmatrix} 0 & 0 & I_p & 0 \\ 0 & 0 & 0 & -I_q \\ -I_p & 0 & 0 & 0 \\ 0 & I_q & 0 & 0 \end{bmatrix} \in \mathfrak{sp}(p+q,\mathbb{R}).$$

(iii) Still in case (3.5.1.3), with reference to the Cartan decomposition

$$\mathfrak{sp}(q,\mathbb{R}) = \mathfrak{u}(q) \oplus \mathfrak{p}$$

of $\mathfrak{sp}(q,\mathbb{R})$, cf. (3.5.1.2), the assignment to

$$\begin{bmatrix} A & B \\ C & -A^{\mathrm{t}}\end{bmatrix} = \begin{bmatrix} A' & B' \\ -B' & A'\end{bmatrix} + \begin{bmatrix} A'' & B'' \\ B'' & -A''\end{bmatrix} \in \mathfrak{sp}(\ell,\mathbb{R})$$

of $\begin{bmatrix} A'+iB' & 0 & iA''+B'' \\ 0 & 0 & 0 \\ -iA''+B'' & 0 & A'-iB' \end{bmatrix} \in \mathfrak{su}(p,q)$ when $p>q$ and of

$$\begin{bmatrix} A'+iB' & iA''+B'' \\ -iA''+B'' & A'-iB' \end{bmatrix} \in \mathfrak{su}(p,q)$$

when $p=q$ where $\begin{bmatrix} A' & B' \\ -B' & A' \end{bmatrix} \in \mathfrak{u}(q)$ and $\begin{bmatrix} A'' & B'' \\ B'' & -A'' \end{bmatrix} \in \mathfrak{p}$ yields a rank preserving H_1-injection

$$\mathfrak{sp}(q,\mathbb{R}) \longrightarrow \mathfrak{su}(p,q). \tag{3.5.1.6}$$

This is an embedding of the kind asserted in Proposition 3.3.2(4); it identifies the two H-elements in the regular case ($p=q$).
(iv) In case (3.5.1.4), the form $\mathcal{B}$, written out as $\mathcal{B}=\mathcal{B}_1+i\mathcal{B}_2+\mathcal{B}_3 j$ with real forms $\mathcal{B}_1$, $\mathcal{B}_2$ and complex form $\mathcal{B}_3$, has $\mathfrak{u}(V,\mathcal{B}_1)=\mathfrak{sp}(2n,\mathbb{R})$, $\mathfrak{u}(V,\mathcal{B}_2)=\mathfrak{so}(2n,2n)$, and $\mathfrak{u}(V,\mathcal{B}_3)=\mathfrak{so}(2n,\mathbb{C})$. Hence $\mathfrak{so}^*(2n)$ is the intersection in $\mathfrak{gl}(4n,\mathbb{R})$ of the Lie subalgebras $\mathfrak{sp}(2n,\mathbb{R})$, $\mathfrak{so}(2n,2n)$, and $\mathfrak{so}(2n,\mathbb{C})$, and the resulting inclusion

$$\mathfrak{so}^*(2n) \longrightarrow \mathfrak{sp}(2n,\mathbb{R}) \tag{3.5.1.7}$$

is an H_1-injection.
(v) Still in case (3.5.1.4), with reference to the Cartan decomposition

$$\mathfrak{sp}(\ell,\mathbb{R}) = \mathfrak{u}(\ell)\oplus\mathfrak{p}$$

of $\mathfrak{sp}(\ell,\mathbb{R})$, cf. (3.5.1.2), the assignment to

$$\begin{bmatrix} A & B \\ C & -A^{\mathrm{t}} \end{bmatrix} = \begin{bmatrix} A' & B' \\ -B' & A' \end{bmatrix} + \begin{bmatrix} A'' & B'' \\ B'' & -A'' \end{bmatrix} \in \mathfrak{sp}(\ell,\mathbb{R})$$

of

$$\begin{bmatrix} U & -V \\ \overline{V} & \overline{U} \end{bmatrix} \in \mathfrak{so}^*(4\ell), \quad U=\begin{bmatrix} A' & iB'' \\ -iB'' & A' \end{bmatrix}, \; V=\begin{bmatrix} -B' & iA'' \\ -iA'' & -B' \end{bmatrix}$$

where $\begin{bmatrix} A' & B' \\ -B' & A' \end{bmatrix} \in \mathfrak{u}(q)$ and $\begin{bmatrix} A'' & B'' \\ B'' & -A'' \end{bmatrix} \in \mathfrak{p}$ yields a rank preserving H_1-injection of $\mathfrak{sp}(\ell,\mathbb{R})$ into $\mathfrak{so}^*(4\ell)$; combination with the obvious injection of $\mathfrak{so}^*(4\ell)$ into $\mathfrak{so}^*(4\ell+2)$, gives a rank preserving H_1-injection of $\mathfrak{sp}(\ell,\mathbb{R})$ into $\mathfrak{so}^*(4\ell+2)$ as well. The resulting rank preserving H_1-morphism

$$\mathfrak{sp}\left(\left[\frac{n}{2}\right],\mathbb{R}\right) \longrightarrow \mathfrak{so}^*(2n), \tag{3.5.1.8}$$

for any $n\geq 2$, is an embedding of the kind asserted in Proposition 3.3.2(4); in the regular case (n even), this embedding identifies the H-element of the source with that of the target.

3.5.2. Explicit H_1-triples. With reference to the notation introduced in (3.5.1.2), for $1\leq k\leq \ell$, consider the embedding $\kappa_k\colon \mathfrak{sp}(1,\mathbb{R})\longrightarrow\mathfrak{sp}(\ell,\mathbb{R})$ given by the assignment to $\begin{bmatrix} a & b \\ c & -a \end{bmatrix} \in \mathfrak{sp}(1,\mathbb{R})$ of $\begin{bmatrix} A_k & B_k \\ C_k & -A_k \end{bmatrix}$ where

$$\begin{aligned} A_k &= \mathrm{diag}(0,\dots,0,a,0,\dots,0) \\ B_k &= \mathrm{diag}(0,\dots,0,b,0,\dots,0) \\ C_k &= \mathrm{diag}(0,\dots,0,c,0,\dots,0), \end{aligned}$$

(a, b, c in the k'th position), and let $e_k = \kappa_k(E) \in \mathfrak{sp}(\ell, \mathbb{R})$, $f_k = \kappa_k(F) \in \mathfrak{sp}(\ell, \mathbb{R})$, $h_k = \kappa_k(H) \in \mathfrak{sp}(\ell, \mathbb{R})$. Each (e_k, f_k, h_k) $(1 \leq k \leq \ell)$ is an H_1-triple in $\mathfrak{sp}(\ell, \mathbb{R})$, cf. Ex. III.3.2 on p. 105/106 of [**79**], and these combine to an H_1-embedding of $\mathfrak{sl}(2, \mathbb{R})^\ell$ into $\mathfrak{sp}(\ell, \mathbb{R})$ of the kind spelled out in Proposition 3.3.2. A corresponding H_1-embedding of $\mathfrak{sl}(2, \mathbb{R})^q$ into $\mathfrak{su}(p, q)$ $(p \geq q)$ results from the corresponding H_1-embedding of $\mathfrak{sl}(2, \mathbb{R})^q$ into $\mathfrak{sp}(q, \mathbb{R})$, combined with the H_1-embedding (3.5.1.6) of $\mathfrak{sp}(q, \mathbb{R})$ into $\mathfrak{su}(p, q)$. Alternatively, noting that $\mathfrak{su}(1, 1)$ is isomorphic to $\mathfrak{sl}(2, \mathbb{R})$, we obtain an obvious embedding of $\mathfrak{sl}(2, \mathbb{R})^q \cong \mathfrak{su}(1, 1)^q$ into $\mathfrak{su}(q, q)$; when $p > q$, this embedding, combined with the obvious embedding of $\mathfrak{su}(q, q)$ into $\mathfrak{su}(p, q)$, yields the requisite embedding of $\mathfrak{sl}(2, \mathbb{R})^q$ into $\mathfrak{su}(p, q)$. Likewise, the composite of the H_1-embedding of $\mathfrak{sl}(2, \mathbb{R})^{[\frac{n}{2}]}$ into $\mathfrak{sp}([\frac{n}{2}], \mathbb{R})$ with the H_1-embedding (3.5.1.8) of $\mathfrak{sp}([\frac{n}{2}], \mathbb{R})$ into $\mathfrak{so}^*(2n)$ yields a corresponding H_1-embedding of $\mathfrak{sl}(2, \mathbb{R})^{[\frac{n}{2}]}$ into $\mathfrak{so}^*(2n)$.

We will say that a matrix X in $\mathfrak{g}$ has *square zero* provided $X^2 = 0$, the square being taken in the algebra $\mathrm{End}_{\mathbb{K}}(V)$ of endomorphisms of the standard representation V of $\mathfrak{g}$. A square zero matrix X in $\mathfrak{g}$ is a nilpotent element of $\mathfrak{g}$ (i. e. $\mathrm{ad}(X)$ is a nilpotent endomorphism of $\mathfrak{g}$) since $X^2 = 0$ entails $(\mathrm{ad}(X))^3 = 0$.

THEOREM 3.5.3. *A nilpotent orbit of a standard simple Lie algebra of hermitian type is pseudoholomorphic if and only if it is generated by a square zero matrix. More precisely, given a standard simple Lie algebra $(\mathfrak{g}, z)$ of hermitian type and split rank r, for $0 \leq t, u \leq r, t + u \leq r$, the pseudoholomorphic nilpotent orbit $\mathcal{O}_{t,u}$ is the smooth connected manifold in $\mathfrak{g}$ which consists of square zero matrices in $\mathfrak{g}$ which have $\mathbb{K}$-rank $t + u$ and $\mathbb{K}$-signature $t - u$, and the $\frac{(r+1)(r+2)}{2}$ pseudoholomorphic nilpotent orbits $\mathcal{O}_{t,u}$ constitute a complete list of all pseudoholomorphic nilpotent orbits.*

PROOF. We will explain in detail the case $\mathfrak{g} = \mathfrak{sp}(\ell, \mathbb{R})$ and will give some hints how the other cases may be reduced to this one. Thus, let $X \in \mathfrak{sp}(\ell, \mathbb{R})$ and suppose that $X^2 = 0$. With reference to the Cartan decomposition of $\mathfrak{sp}(\ell, \mathbb{R})$, cf. (3.5.1.2), write

$$(3.5.3.1) \qquad X = \begin{bmatrix} A & B \\ -B & A \end{bmatrix} + \begin{bmatrix} S & T \\ T & -S \end{bmatrix}, \quad A, B, S, T \in \mathrm{M}_{\ell,\ell}(\mathbb{R})$$

where A is skew-symmetric and B, S, T are symmetric. Since every element in $\mathfrak{u}(\ell)$ lies in the Lie algebra of a maximal torus, we may assume that $A = 0$ and that B is a diagonal matrix. Then a straightforward calculation shows that $X^2 = 0$ implies that S and T are diagonal matrices as well, and that $B^2 = S^2 + T^2$. Hence $X \in \mathfrak{g} = \mathfrak{sp}(\ell, \mathbb{R})$ is determined by its projection $\begin{bmatrix} S & T \\ T & -S \end{bmatrix}$ to $\mathfrak{p} = S^2_{\mathbb{R}}[\mathbb{R}^\ell] \oplus S^2_{\mathbb{R}}[\mathbb{R}^\ell]$, up to the signs in the entries of B, and a choice of signs corresponds precisely to a choice of signature for the hermitian form $\mathcal{B}_X$, once the rank is fixed.

Consider the subset $\widetilde{\mathcal{O}}_{t,u}$ of $\mathfrak{g}$ which consists of square zero matrices in $\mathfrak{g}$ which have rank $t+u$ and signature $t-u$. Inspection shows that $X = e_1 + \cdots + e_t - e_{t+1} - \cdots - e_{t+u}$, cf. Proposition 3.3.2 and Subsection 3.5.2, is a square zero matrix in $\mathfrak{g}$. Hence the pseudoholomorphic nilpotent orbit $\mathcal{O}_{t,u}$ is contained in $\widetilde{\mathcal{O}}_{t,u}$. Moreover, for $X \in \widetilde{\mathcal{O}}_{t,u}$, written out as in (3.5.3.1), after diagonalization so that $A = 0$ and so that B is a diagonal matrix, the signature being constant (and equal to $t - u$) amounts to making a possible choice of signs for B. Consequently the restriction to $\widetilde{\mathcal{O}}_{t,u}$ of the projection from $\mathfrak{g} = \mathfrak{k} \oplus \mathfrak{p}$ to $\mathfrak{p}$ is a bijection onto its image in $\mathfrak{p}$. We will show shortly that $\widetilde{\mathcal{O}}_{t,u}$ coincides with $\mathcal{O}_{t,u}$.

In view of the classification of real adjoint orbits by *types* [**7**], nilpotent orbits are classified by sums of *indecomposable (nilpotent) types*, where a(n indecomposable) "type" is a generalized Jordan normal form (Jordan block); cf. also (9.3) in [**9**]. Under our circumstances, this means the following: Given a nilpotent orbit $\mathcal{O}$ of $\mathfrak{g}$ and $X \in \mathcal{O}$, with reference to the operator X on $V = \mathbb{R}^{2\ell}$, the space V decomposes into a sum $V = V_1 \oplus \cdots \oplus V_\tau$ of X-modules V_j which are mutually isotropic with respect to the symplectic form ω (say) on V, that is, $\omega(V_j, V_{j'}) = 0$ for $j \neq j'$, and which are either X-*indecomposable* or are decomposed into two indecomposable X-stable subspaces of the same dimension, and each V_j which is X-indecomposable, i. e. not decomposed into two X-stable subspaces of the same dimension, carries as an additional ingredient a sign $\pm$ which encapsulates the behaviour of the hermitian form $\mathcal{B}_X$. A space of the kind $\widetilde{\mathcal{O}}_{t,u}$ corresponds to a sum of indecomposable types of the kind

$$V = \Delta_1^+(0) \oplus \cdots \oplus \Delta_1^+(0) \oplus \Delta_1^-(0) \oplus \cdots \oplus \Delta_1^-(0) \oplus \Delta_0(0,0) \oplus \cdots \oplus \Delta_0(0,0)$$

involving t copies of $\Delta_1^+(0)$, u copies of $\Delta_1^-(0)$, and $r - t - u$ copies of $\Delta_0(0,0)$; here the notation is that in [**7**], each summand is a real symplectic vector space of dimension two, for $X \in \widetilde{\mathcal{O}}_{t,u}$ the form $\mathcal{B}_X$ is positive on each copy of $\Delta_1^+(0)$, negative on each copy of $\Delta_1^+(0)$, and zero on each copy of $\Delta_0(0,0)$. This shows in particular that each $\widetilde{\mathcal{O}}_{t,u}$ is a nilpotent orbit which necessarily coincides with $\mathcal{O}_{t,u}$ and that, when $X \in \mathfrak{g}$ has square zero, the only possible indecomposable types in the decomposition of V are $\Delta_1^+(0)$, $\Delta_1^-(0)$, and $\Delta_0(0,0)$. Indeed, when a higher type occurs, the operator X^2 is non-zero on this type.

The proof is then completed by the observation that, if a nilpotent orbit $\mathcal{O}$ involves a type Δ of a kind different from those spelled out above, the projection from $\mathcal{O}$ to $\mathfrak{p}$ is no longer injective. Indeed, it suffices to consider the special case where V is the underlying space of such a higher type Δ, that is, where $\mathfrak{g} = \mathfrak{sp}(V) = \mathfrak{sp}(m, \mathbb{R})$ for some $1 < m \leq \ell$. There are only two kinds of types:
1) $X^{2m-1} \neq 0$. In this case, $\mathcal{O}$ is one of the two *real* regular nilpotent orbits in $\mathfrak{sp}(m, \mathbb{R})$. However, the dimension of each of the two regular orbits is plainly strictly larger than that of the constituent $\mathfrak{p}$ in the Cartan decomposition of $\mathfrak{sp}(m, \mathbb{R})$ unless $m = 1$ but the case $m = 1$ is excluded here since we suppose that the type Δ under consideration is different from those spelled out above. Hence the projection from $\mathcal{O}$ to $\mathfrak{p}$ cannot then be injective. Indeed, $\dim(\mathfrak{sp}(m, \mathbb{R})) = 2m^2 + m$ and $\dim(\mathcal{O}) = 2m^2$ whereas $\dim(\mathfrak{p}) = m^2 + m$. In the literature, this type is written as $\Delta_k^{\pm}(0)$, with $k = 2m - 1$.
2) As an X-module, V decomposes into a sum $V = V' \oplus V''$ of two Lagrangian subspaces $V' \cong \mathbb{R}^m$ and $V'' \cong \mathbb{R}^m$, each of which is X-indecomposable, in such a way that the orbit $\mathcal{O}$ arises from the (nonnegative or from the nonpositive) regular nilpotent orbit of $\mathfrak{gl}(m, \mathbb{R})$ in the following fashion: Consider the canonical injection of $\mathfrak{gl}(m, \mathbb{R})$ into of $\mathfrak{sp}(m, \mathbb{R})$ which sends a matrix $A \in \mathfrak{gl}(m, \mathbb{R})$ to the matrix $\widehat{A} = \begin{bmatrix} A & 0 \\ 0 & -A^t \end{bmatrix} \in \mathfrak{sp}(m, \mathbb{R})$. Up to a choice of basis in V' and V'', the endomorphism X of V is now of the kind $X = \widehat{A}$, where A is a regular nilpotent element of $\mathfrak{gl}(m, \mathbb{R})$, and it does not matter whether A is taken from the nonnegative or nonpositive regular nilpotent orbit of $\mathfrak{gl}(m, \mathbb{R})$. For example, A could be the ordinary Jordan

normal form of a nilpotent rank $m-1$ matrix in $\mathfrak{gl}(m,\mathbb{R})$. In the literature, this type is written as $\Delta_k(0,0)$, with $k=m-1$. (It yields new types only for k even.) To verify that the projection from $\mathcal{O}$ to the constituent $\mathfrak{p}$ of the Cartan decomposition $\mathfrak{sp}(m,\mathbb{R})=\mathfrak{u}(m)\oplus\mathfrak{p}$ is no longer injective, it suffices to prove that the projection from one of the two regular nilpotent orbits $\mathcal{O}'$ (say) of $\mathfrak{gl}(m,\mathbb{R})$ to the constituent $\mathfrak{p}'\cong \mathrm{S}^2_{\mathbb{R}}[\mathbb{R}^m]$ of the Cartan decomposition $\mathfrak{gl}(m,\mathbb{R})=\mathfrak{so}(m)\oplus\mathfrak{p}'$ is not injective. However, $\mathfrak{gl}(m,\mathbb{R})$ is reductive and not simple, and $\mathcal{O}'$ is actually an orbit in $\mathfrak{sl}(m,\mathbb{R})$ whence $\mathcal{O}'$ projects to the constituent $\mathfrak{p}''\subseteq\mathfrak{p}'$ of the Cartan decomposition $\mathfrak{sl}(m,\mathbb{R})=\mathfrak{so}(m)\oplus\mathfrak{p}''$; notice that $\mathfrak{p}''$ consists of the symmetric trace zero $(m\times m)$-matrices. Thus, when $m\geq 3$,

$$\dim(\mathcal{O}')=m^2-m>\frac{1}{2}m(m+1)-1=\dim(\mathfrak{p}'')$$

whence the projection from $\mathcal{O}'$ to $\mathfrak{p}''$ cannot be injective. The case $m=2$ (i. e. the type $\Delta_1(0,0)$) is already covered by the type $\Delta_1^+(0)\oplus\Delta_1^-(0)$ examined earlier.

A similar reasoning establishes the claim in the other standard cases. Alternatively, the embeddings (3.5.1.4) of $\mathfrak{su}(p,q)$ into $\mathfrak{sp}(p+q,\mathbb{R})$ and (3.5.1.6) of $\mathfrak{so}^*(2n)$ into $\mathfrak{sp}(2n,\mathbb{R})$ are compatible with the Cartan decompositions and H_1-embeddings. By means of these embeddings, we may reduce the case of $\mathfrak{su}(p,q)$ and that of $\mathfrak{so}^*(2n)$ to that of $\mathfrak{sp}(\ell,\mathbb{R})$ (for suitable ℓ). More precisely: For $\mathfrak{su}(p,q)$, we have the chain of H_1-injections

$$\mathfrak{sp}(q,\mathbb{R})\subseteq\mathfrak{su}(p,q)\subseteq\mathfrak{sp}(p+q,\mathbb{R}),$$

and things may be arranged in such a way that, cf. Proposition 3.3.2 above, a complete system

$$(e_1,f_1,h_1),\ldots,(e_q,f_q,h_q)$$

of H_1-triples for $\mathfrak{sp}(q,\mathbb{R})$ constitutes a complete system of H_1-triples for $\mathfrak{su}(p,q)$ and that, together with suitable additional H_1-triples

$$(e_{q+1},f_{q+1},h_{q+1}),\ldots,(e_{p+q},f_{p+q},h_{p+q}),$$

the H_1-triples

$$(e_1,f_1,h_1),\ldots,(e_{p+q},f_{p+q},h_{p+q})$$

constitute a complete system of H_1-triples for $\mathfrak{sp}(p+q,\mathbb{R})$. Then, for $0\leq t,u\leq q, t+u\leq q$, the intersection $\mathcal{O}_{t,u}\cap\mathfrak{su}(p,q)$ of the corresponding pseudoholomorphic nilpotent orbit $\mathcal{O}_{t,u}$ for $\mathfrak{sp}(p+q,\mathbb{R})$ is the corresponding pseudoholomorphic nilpotent orbit for $\mathfrak{su}(p,q)$. The same kind of reasoning applies to $\mathfrak{so}^*(2n)$. We leave the details to the reader. Furthermore, the statement referring to the closures of the orbits $\mathcal{O}_{t,u}$ is a consequence of the observation that passing to the closure amounts to dropping summands in the decomposition of types. □

As noted by Springer-Steinberg [**87**], nilpotent orbits in $\mathfrak{sp}(\ell,\mathbb{C})$ arise as the intersections of nilpotent orbits in $\mathfrak{sl}(2\ell,\mathbb{C})$ with $\mathfrak{sp}(\ell,\mathbb{C})$. Theorem 3.5.3 entails that, given a nilpotent orbit $\widetilde{\mathcal{O}}$ in $\mathfrak{sp}(\ell,\mathbb{C})$ arising as the complexification of a pseudoholomorphic nilpotent orbit in $\mathfrak{sp}(\ell,\mathbb{R})$, the intersection $\widetilde{\mathcal{O}}\cap\mathfrak{sp}(\ell,\mathbb{R})$ decomposes into connected components which are distinguished by the signature of the form $\mathcal{B}_X$ for $X\in\widetilde{\mathcal{O}}\cap\mathfrak{sp}(\ell,\mathbb{R})$. The same kind of remark applies to $\mathfrak{su}(p,q)$ and $\mathfrak{so}^*(2n)$. In particular, in the standard cases, we can thus single out the holomorphic nilpotent orbits among the pseudoholomorphic ones. We will say that $X\in\mathfrak{u}(V,\mathcal{B})$ is *non-negative* (*non-positive*) provided the hermitian form $\mathcal{B}_X$ is non-negative (non-positive); plainly, the property of X being non-negative (non-positive) depends only on the $\mathrm{U}(V,\mathcal{B})^0$-orbit of X, and we can talk about *non-negative* (*non-positive*) $\mathrm{U}(V,\mathcal{B})^0$-orbits. The non-negative elements X in $\mathfrak{u}(V,\mathcal{B})$ constitute a (real) cone in $\mathfrak{u}(V,\mathcal{B})$ which is invariant under $\mathrm{U}(V,\mathcal{B})^0$ (and the same is true, of course, of the non-positive elements).

THEOREM 3.5.4. *Given a standard simple Lie algebra $(\mathfrak{g},z)$ of hermitian type and real rank r, for $1 \le s \le r$, the holomorphic nilpotent orbit $\mathcal{O}_s$ $(= \mathcal{O}_{s,0})$ is the (connected) smooth manifold of non-negative rank s nilpotent matrices in $\mathfrak{g}$ (real rank s for $\mathfrak{g} = \mathfrak{sp}(r,\mathbb{R})$; complex rank s for $\mathfrak{g} = \mathfrak{su}(p,q)$, where $p \ge q = r$; quaternionic rank s for $\mathfrak{g} = \mathfrak{so}^*(2n)$, where $r = [\frac{n}{2}]$). Furthermore, as a real semi-algebraic set in $\mathfrak{g}$, the closure $\overline{\mathcal{O}_s}$ is the space of non-negative nilpotent matrices in $\mathfrak{g}$ which have rank at most s.*

PROOF. By Theorem 3.5.3, the holomorphic nilpotent orbit $\mathcal{O}_s$ $(1 \le s \le r)$ is the smooth connected manifold of non-negative rank s nilpotent square zero matrices in $\mathfrak{g}$. Theorem 5.4.1 below implies that a non-negative nilpotent matrix generates a holomorphic nilpotent orbit and is therefore of square zero, and the "Furthermore" statement is implied by that theorem as well. □

Thus, in the definition of the 2-form $\mathcal{B}_X$, cf. (3.5.1.1), the sign has been adjusted in such a way that $\mathcal{B}_X$ being non-negative corresponds to the holomorphicity of the orbit generated by X. Explicit equations and inequalities describing $\overline{\mathcal{O}_s}$ as a real semi-algebraic set in $\mathfrak{g}$ will be given in Remark 5.4.2.

In the standard cases, the complex analytic structures of the holomorphic nilpotent orbits and their closures are elucidated by the following result which, apart from the last statement referring to the normality of the resulting complex varieties, makes explicit the statement of Theorem 3.3.11 above for the standard cases and in particular spells out the $K^{\mathbb{C}}$-stratification.

Recall that, with reference to the Cartan decomposition $\mathfrak{g} = \mathfrak{k} \oplus \mathfrak{p}$, $\mathfrak{p}^+ = \mathrm{S}^2_{\mathbb{C}}[\mathbb{C}^r]$ for $\mathfrak{g} = \mathfrak{sp}(r,\mathbb{R})$, $\mathfrak{p}^+ = \mathrm{M}_{q,p}(\mathbb{C})$ for $\mathfrak{g} = \mathfrak{su}(p,q)$ $(p \ge q = r)$, $\mathfrak{p}^+ = \Lambda^2_{\mathbb{C}}[\mathbb{C}^n]$ for $\mathfrak{g} = \mathfrak{so}^*(2n,\mathbb{R})$ $(r = [\frac{n}{2}])$ where r denotes the real rank. For $1 \le s < r$, let V_s be the smooth complex affine variety in $\mathfrak{p}^+$ which consists of complex matrices in $\mathfrak{p}^+$ having rank s for $\mathfrak{g} = \mathfrak{sp}(r,\mathbb{R})$ and $\mathfrak{g} = \mathfrak{su}(p,q)$ and rank $2s$ for $\mathfrak{g} = \mathfrak{so}^*(2n)$; likewise, let $V_{\le s}$ be the complex affine determinantal variety in $\mathfrak{p}^+$ which consists of complex matrices in $\mathfrak{p}^+$ having rank at most s for $\mathfrak{g} = \mathfrak{sp}(r,\mathbb{R})$ and $\mathfrak{g} = \mathfrak{su}(p,q)$ and rank at most $2s$ for $\mathfrak{g} = \mathfrak{so}^*(2n)$.

THEOREM 3.5.5. *For a standard simple rank r Lie algebra of hermitian type $(\mathfrak{g},z)$, under the projection from $\mathfrak{g} = \mathfrak{k} \oplus \mathfrak{p}$ to $\mathfrak{p}$, followed by the complex linear isomorphism from $\mathfrak{p}$ to $\mathfrak{p}^+$, the closure $\overline{\mathcal{O}_r}$ of the principal holomorphic nilpotent orbit is identified with $\mathfrak{p}^+$ and, for $1 \le s < r$, $\mathcal{O}_s$ is mapped diffeomorphically onto the smooth affine complex subvariety V_s of $\mathfrak{p}^+$. Likewise, for $1 \le s < r$, as a complex analytic space, $\overline{\mathcal{O}_s}$ is identified with the complex affine determinantal variety $V_{\le s}$ in $\mathfrak{p}^+$. An explicit system of equations defining $\overline{\mathcal{O}_s}$ as an affine complex variety is thus given by the determinantal equations which in the first two cases say that all $((s+1)\times(s+1))$-minors are zero and which in the third case say that all $(2(s+1)\times 2(s+1))$-minors are zero. As a complex variety, each $\overline{\mathcal{O}_s}$ is normal.* □

Determinantal varieties have been studied in the literature; see e. g. [**6**], [**26**]; they are known to be normal [**6**] (Theorem 2.11), [**26**]. Thus, at the present stage, the proof of Theorem 3.5.5 is complete. A result more general than Theorem 3.5.5 will be given as Theorem 5.3.3 below which will, in particular, entail the normality of the closures of the strata and hence the normality of complex determinantal varieties. The reasoning aimed at unravelling the complex analytic structures of the categorical quotients $W_J//H^{\mathbb{C}}$ in the proof of Theorem 5.3.3 below establishes the claim of Theorem 3.5.5 as well.

REMARK 3.5.6. The proof of Theorem 3.5.3 is independent of those of Theorems 3.3.3 and 3.3.11 and thus yields another proof of the statement of Theorem 3.3.3 and of that of Theorem 3.3.11 for the special case where $(\mathfrak{g}, z)$ is a standard simple rank r Lie algebra of hermitian type.

3.6. THE LIE ALGEBRA $\mathfrak{g} = \mathfrak{so}(2,q)$, $q \geq 2$. This is the remaining classical case. We note that $\mathfrak{so}(2,1) \cong \mathfrak{sp}(1,\mathbb{R})$, $\mathfrak{so}(2,3) \cong \mathfrak{sp}(2,\mathbb{R})$, $\mathfrak{so}(2,4) \cong \mathfrak{su}(2,2)$, $\mathfrak{so}(2,6) \cong \mathfrak{so}^*(8)$, and that $\mathfrak{so}(2,10)$ arises from the real Cayley division algebra in a similar fashion; see Section 8 below. Recall that

$$\mathfrak{g} = \mathfrak{so}(2,q) = \left\{ \begin{bmatrix} A & B \\ B^{\mathrm{t}} & D \end{bmatrix}; A \in \mathfrak{so}(2,\mathbb{R}), D \in \mathfrak{so}(q,\mathbb{R}), B \in \mathrm{M}_{q,2}(\mathbb{R}) \right\}.$$

Its split rank r equals 2, and $z = \begin{bmatrix} J & 0 \\ 0 & 0 \end{bmatrix}$ where $J = \begin{bmatrix} 0 & 1 \\ -1 & 0 \end{bmatrix}$ is an H-element, so that $(\mathfrak{g}, z)$ is a simple Lie algebra of hermitian type for $q \geq 3$. The constituents $\mathfrak{k}$ and $\mathfrak{p}$ of its Cartan decomposition have the form $\mathfrak{k} = \mathfrak{so}(2,\mathbb{R}) \oplus \mathfrak{so}(q,\mathbb{R})$ and $\mathfrak{p} = \left\{ \begin{bmatrix} 0 & B \\ B^{\mathrm{t}} & 0 \end{bmatrix} \right\} \cong \mathrm{M}_{q,2}(\mathbb{R})$ and, for $x = \begin{bmatrix} 0 & B \\ B^{\mathrm{t}} & 0 \end{bmatrix} \in \mathfrak{p}$, the value $J_z(x)$ under the complex operator J_z on $\mathfrak{p}$ is easily seen to be given by matrix multiplication by J on B from the left.

The Lie algebra $\mathfrak{so}(2,2)$ is generated by

$$X = \begin{bmatrix} J & 0 \\ 0 & J \end{bmatrix}, \quad A_1 = \begin{bmatrix} 0 & 0 & 1 & 0 \\ 0 & 0 & 0 & -1 \\ 1 & 0 & 0 & 0 \\ 0 & -1 & 0 & 0 \end{bmatrix}, \quad A_2 = \begin{bmatrix} 0 & 0 & 0 & 1 \\ 0 & 0 & 1 & 0 \\ 0 & 1 & 0 & 0 \\ 1 & 0 & 0 & 0 \end{bmatrix}$$

$$Y = \begin{bmatrix} J & 0 \\ 0 & -J \end{bmatrix}, \quad B_1 = \begin{bmatrix} 0 & 0 & 0 & 1 \\ 0 & 0 & -1 & 0 \\ 0 & -1 & 0 & 0 \\ 1 & 0 & 0 & 0 \end{bmatrix}, \quad B_2 = \begin{bmatrix} 0 & 0 & 1 & 0 \\ 0 & 0 & 0 & 1 \\ 1 & 0 & 0 & 0 \\ 0 & 1 & 0 & 0 \end{bmatrix},$$

these matrices satisfy the relations

$$\begin{aligned} &[A_1, A_2] = 2X, \quad [X, A_1] = -2A_2, \quad [X, A_2] = 2A_1, \\ &[B_1, B_2] = 2Y, \quad [Y, B_1] = -2B_2, \quad [Y, B_2] = 2B_1, \end{aligned}$$

and (X, A_1, A_2) and (Y, B_1, B_2) each generate a copy of $\mathfrak{sl}(2,\mathbb{R}) \cong \mathfrak{so}(2,1)$ in $\mathfrak{so}(2,2)$ in such a way that $\mathfrak{so}(2,2) = \mathfrak{so}(2,1) \oplus \mathfrak{so}(2,1)$. Let

$$\begin{aligned} e_1 &= \frac{1}{2}(A_2 + X), \quad f_1 = \frac{1}{2}(A_2 - X), \quad h_1 = A_1, \\ e_2 &= \frac{1}{2}(B_2 + Y), \quad f_2 = \frac{1}{2}(B_2 - Y), \quad h_2 = B_1. \end{aligned}$$

These lie in $\mathfrak{so}(2,2)$ and, via the standard embedding of $\mathfrak{so}(2,2)$ into $\mathfrak{so}(2,q)$ for $q \geq 3$, we view $e_1, f_1, h_1, e_2, f_2, h_2$ as elements of $\mathfrak{so}(2,q)$; inspection shows that (e_1, f_1, h_1) and (e_2, f_2, h_2) are holomorphic $\mathfrak{sl}(2)$-triples in $\mathfrak{so}(2,q)$, and these combine to an H_1-embedding of $\mathfrak{sl}(2,\mathbb{R})^2$ into $\mathfrak{so}(2,q)$ of the kind spelled out in Proposition 3.3.2. By Theorem 3.3.3, with the notation $G = \mathrm{SO}(2,q)^0$, for $q \geq 3$, the orbits

$$\begin{gathered}\mathcal{O}_{0,0} = \{0\},\ \mathcal{O}_{1,0} = Ge_1,\ \mathcal{O}_{0,1} = G(-e_1),\\ \mathcal{O}_{2,0} = G(e_1+e_2),\ \mathcal{O}_{1,1} = G(e_1-e_2),\ \mathcal{O}_{0,2} = G(e_1+e_2),\end{gathered} \tag{3.6.1}$$

are pseudoholomorphic, $\mathcal{O}_{1,0}$, $\mathcal{O}_{2,0}$ are holomorphic, $\mathcal{O}_{0,1}$, $\mathcal{O}_{0,2}$ are antiholomorphic, and $\mathcal{O}_{1,1}$ is neither holomorphic nor antiholomorphic. Recall that $\mathrm{SO}(2,q)$ and $\mathrm{O}(2,q)$ have two and four connected components, respectively, and that $\mathrm{O}(2,q)^0 = \mathrm{SO}(2,q)^0$. Notice that a matrix X in $\mathfrak{g}$ with $X^3 = 0$ satisfies $(\mathrm{ad}(X))^4 = 0$ and is therefore a nilpotent element of $\mathfrak{g}$.

THEOREM 3.6.2. *For $q \geq 3$, (3.6.1) constitutes a complete list of the pseudoholomorphic nilpotent $\mathrm{SO}(2,q)^0$-orbits in $\mathfrak{so}(2,q)$. Furthermore:*
(i) *The orbits $\mathcal{O}_{1,0}$ and $\mathcal{O}_{0,1}$ are the two connected components of the single $\mathrm{SO}(2,q)$-orbit (and hence $\mathrm{O}(2,q)$-orbit) in $\mathfrak{so}(2,q)$ which consists of square zero matrices in $\mathfrak{so}(2,q)$.*
(ii) *The three $\mathrm{SO}(2,q)^0$-orbits $\mathcal{O}_{2,0}$, $\mathcal{O}_{1,1}$, $\mathcal{O}_{0,2}$ are the connected components of the smooth manifold of matrices X in $\mathfrak{so}(2,q)$ satisfying $X^3 = 0$ in such a way that, as $\mathrm{SO}(2,q)$-orbits as well as as $\mathrm{O}(2,q)$-orbits, $\mathcal{O}_{2,0}$ and $\mathcal{O}_{0,2}$ are the two connected components of one orbit and $\mathcal{O}_{1,1}$ is the other orbit.*
(iii) *For $q \geq 2$, the orbits $\mathcal{O}_{1,0}$ and $\mathcal{O}_{0,1}$ are the two connected components of the smooth manifold of rank two nilpotent matrices $\begin{bmatrix} A & B \\ B^{\mathrm{t}} & D\end{bmatrix}$ in $\mathfrak{so}(2,q)$ which, when B is written as $B = \begin{bmatrix} x_1 & x_2 & \cdots & x_q \\ y_1 & y_2 & \cdots & y_q\end{bmatrix}$, satisfy the equations*

$$\begin{aligned} x_1^2 - y_1^2 + x_2^2 - y_2^2 + \cdots + x_q^2 - y_q^2 &= 0\\ x_1y_1 + x_2y_2 + \cdots + x_qy_q &= 0.\end{aligned}$$

PROOF. Statements (i) and (ii) are established by a reasoning involving the decomposition of the type of a nilpotent orbit into a sum of indecomposable $\mathrm{O}(2,q)$-types [**7**], similar to that given in the proof of Theorem 3.5.3, together with a closer look at the decomposition of the resulting $\mathrm{O}(2,q)$-orbits into $\mathrm{SO}(2,q)$- and $\mathrm{SO}(2,q)^0$-orbits; we leave the details to the reader. Statement (iii) is a consequence of Theorem 3.6.3 below. It may as well be deduced from the discussion in Section 6 below. □

The next result makes explicit the statement of Theorem 3.3.11 above for $\mathfrak{g} = \mathfrak{so}(2,q)$.

THEOREM 3.6.3. *Under the projection from $\mathfrak{g} = \mathfrak{k} \oplus \mathfrak{p}$ to $\mathfrak{p}$, followed by the canonical isomorphism from $\mathfrak{p}$ to $\mathfrak{p}^+$, the closure $\overline{\mathcal{O}_{2,0}}$ of the principal holomorphic nilpotent orbit $\mathcal{O}_{2,0}$ is identified with $\mathfrak{p}^+ \cong \mathbb{C}^q$, and the closure $\overline{\mathcal{O}_{1,0}}$ of $\mathcal{O}_{1,0}$ amounts to the smooth complex affine quadric in $\mathbb{C}^q$ given by the equation $w_1^2 + w_2^2 + \cdots + w_q^2 = 0$. Furthermore, as a complex analytic space, $\overline{\mathcal{O}_{1,0}}$ is normal, and so is, of course, $\overline{\mathcal{O}_{2,0}}$ since it is an affine space.*

PROOF. The first statement is established by inspection. See also Remark 6.3 below. Being an affine quadric, the normality of $\overline{\mathcal{O}_{1,0}}$ is immediate. □

REMARK 3.6.4. The proof of Theorem 3.6.2 is independent of those of Theorems 3.3.3 and 3.3.11 and thus yields proofs of the statements of these Theorems for the present special case as well.

Since $\mathfrak{so}(2,3)$ is isomorphic to $\mathfrak{sp}(2,\mathbb{R})$, for $q \geq 3$, the standard inclusion of $\mathfrak{so}(2,3)$ into $\mathfrak{so}(2,q)$ amounts to an injection

$$\mathfrak{sp}(2,\mathbb{R}) \longrightarrow \mathfrak{so}(2,q). \tag{3.6.5}$$

This is an H_1-embedding of the kind asserted in Proposition 3.3.2(4); it identifies the H-element of the source with that of the target.

3.7. THE COMPLEX ANALYTIC STRATIFIED KÄHLER STRUCTURE. We will now show that the closure of any holomorphic nilpotent orbit $\mathcal{O}$ in a semisimple Lie algebra of hermitian type carries the structure of a complex analytic stratified Kähler space which, by virtue of Theorems 3.5.5, 3.6.3, 5.3.3, 8.4.1, 8.4.2, is actually normal. Thus, let $(\mathfrak{g}, z)$ be a semisimple Lie algebra of hermitian type, and let $\mathcal{O}$ be a holomorphic nilpotent orbit. Its closure $N = \overline{\mathcal{O}}$ in $\mathfrak{g}$ is a union of finitely many holomorphic nilpotent orbits and, as we have seen above, the restriction to N of the projection from $\mathfrak{g}$ to the constituent $\mathfrak{p}$ of the Cartan decomposition of $\mathfrak{g}$ is an injective map, even though the projection from $\mathfrak{g}$ to $\mathfrak{p}$ is not injective; this injection turns N into a complex analytic space in such a way that each G-orbit in N is a smooth complex manifold. Independently of the structure results obtained above, the decomposition of N into G-orbits may be shown to be a stratification, since $\mathfrak{g}$ is semisimple; see [**60**] (Theorem 2.23). Furthermore, the *Tarski-Seidenberg* Theorem entails that, as a subset of $\mathfrak{g}$, N is real semi-algebraic.

Upon identifying $\mathfrak{g}$ with its dual $\mathfrak{g}^*$ by means of an appropriate positive multiple of the Killing form we view N as the closure of a (nilpotent) *coadjoint* orbit (i. e. orbit in $\mathfrak{g}^*$). At this stage any positive multiple of the Killing form will do; in Section 5 below a more specific choice will be made; see also (3.2.2) above. Restricting the algebra $C^\infty(\mathfrak{g}^*)$ of ordinary smooth functions on $\mathfrak{g}^*$ to N yields a smooth structure $C^\infty(N)$ on N; a smooth structure of this kind is usually referred to as an algebra of *Whitney smooth functions*. The existence of smooth partitions of unity on $\mathfrak{g}^*$ entails that this smooth structure is fine. Furthermore, since N is the closure of a nilpotent orbit, the Lie-Poisson structure $\{\cdot,\cdot\}$ on $C^\infty(\mathfrak{g}^*)$ passes to a Poisson bracket $\{\cdot,\cdot\}$ on $C^\infty(N)$ which, on each stratum of N (i. e. coadjoint orbit contained in N), restricts to the ordinary symplectic Poisson algebra. Thus $(C^\infty(N), \{\cdot,\cdot\})$ *turns* N *into a stratified symplectic space.*

To spell out the requisite complex analytic stratified Kähler polarization, we note that, as a module over the algebra $C^\infty(\mathfrak{g}^*,\mathbb{C})$, the module $\Omega^1(\mathfrak{g}^*,\mathbb{C})$ of 1-forms may be written in the form $C^\infty(\mathfrak{g}^*,\mathbb{C}) \otimes_{\mathbb{C}} \mathfrak{g}^{\mathbb{C}}$. Hence the decomposition $\mathfrak{g}^{\mathbb{C}} = \mathfrak{k}^{\mathbb{C}} \oplus \mathfrak{p}^+ \oplus \mathfrak{p}^-$ entails a direct sum decomposition

$$\Omega^1(\mathfrak{g}^*,\mathbb{C}) = C^\infty(\mathfrak{g}^*,\mathbb{C}) \otimes_{\mathbb{C}} \mathfrak{k}^{\mathbb{C}} \oplus C^\infty(\mathfrak{g}^*,\mathbb{C}) \otimes_{\mathbb{C}} \mathfrak{p}^+ \oplus C^\infty(\mathfrak{g}^*,\mathbb{C}) \otimes_{\mathbb{C}} \mathfrak{p}^-$$

as a module over the algebra $C^\infty(\mathfrak{g}^*,\mathbb{C})$. The Lie-Rinehart structure on $\Omega^1(\mathfrak{g}^*,\mathbb{C})$ coming from the Poisson structure on $C^\infty(\mathfrak{g}^*,\mathbb{C})$ (explained in Section 1 above) is actually a "crossed product structure", cf. [**29**]. Since $\mathfrak{k}^{\mathbb{C}}$ is a (complex) Lie algebra, the constituent $C^\infty(\mathfrak{g}^*,\mathbb{C}) \otimes_{\mathbb{C}} \mathfrak{k}^{\mathbb{C}}$ is closed under the Lie bracket in $\Omega^1(\mathfrak{g}^*,\mathbb{C})$ and hence inherits a Lie-Rinehart structure, a crossed product structure as well. We have already observed that the constituents $\mathfrak{p}^+$ and $\mathfrak{p}^-$ are known to be abelian subalgebras of $\mathfrak{g}^{\mathbb{C}}$, cf. **[24]** (p. 313), **[79]** (p. 55). Hence the constituents $C^\infty(\mathfrak{g}^*, \mathbb{C}) \otimes_{\mathbb{C}} \mathfrak{p}^+$ and $C^\infty(\mathfrak{g}^*, \mathbb{C}) \otimes_{\mathbb{C}} \mathfrak{p}^-$ are closed under the Lie bracket in $\Omega^1(\mathfrak{g}^*, \mathbb{C})$ and inherit Lie-Rinehart structures; these are crossed product structures again. Even though $\mathfrak{p}^+$ and $\mathfrak{p}^-$ are abelian, the brackets on $C^\infty(\mathfrak{g}^*, \mathbb{C}) \otimes_{\mathbb{C}} \mathfrak{p}^+$ and $C^\infty(\mathfrak{g}^*, \mathbb{C}) \otimes_{\mathbb{C}} \mathfrak{p}^-$ are non-trivial since they are induced from brackets of the kind $[u, v]$ where $u \in \mathfrak{p}^+$ and $v \in \mathfrak{g}^{\mathbb{C}}$.

Write $A(N) = C^\infty(N, \mathbb{C})$, and let I be the ideal of smooth functions on $\mathfrak{g}^*$ which vanish on N. By the general theory of Kähler differentials, the projection map from $C^\infty(\mathfrak{g}^*, \mathbb{C})$ to $A(N)$ gives rise to the exact sequence

$$I/I^2 \xrightarrow{\delta} A(N) \otimes_{C^\infty(\mathfrak{g}^*, \mathbb{C})} \Omega^1(\mathfrak{g}^*, \mathbb{C}) \to \Omega^1(N, \mathbb{C}) \to 0$$

of $A(N)$-modules where δ is induced by the association $a \mapsto 1 \otimes da$ $(a \in I)$. Furthermore, as an $A(N)$-module, $A(N) \otimes_{C^\infty(\mathfrak{g}^*, \mathbb{C})} \Omega^1(\mathfrak{g}^*, \mathbb{C})$ may plainly be written as

$$A(N) \otimes_{C^\infty(\mathfrak{g}^*, \mathbb{C})} \Omega^1(\mathfrak{g}^*, \mathbb{C}) \cong A(N) \otimes_{\mathbb{C}} \mathfrak{k}^{\mathbb{C}} \oplus A(N) \otimes_{\mathbb{C}} \mathfrak{p}^+ \oplus A(N) \otimes_{\mathbb{C}} \mathfrak{p}^-.$$

Let $P \subseteq \Omega^1(N, \mathbb{C})$ be the image of $A(N) \otimes_{\mathbb{C}} \mathfrak{p}^+$ in $\Omega^1(N, \mathbb{C})$. This is a $(\mathbb{C}, A(N))$-Lie subalgebra of $\Omega^1(N, \mathbb{C})_{\{\cdot,\cdot\}}$ which, as an $A(N)$-module, is generated by the differentials dw where $w \in \mathfrak{p}^+$; here the elements of $\mathfrak{g}^{\mathbb{C}}$ are viewed as functions on $(\mathfrak{g}^{\mathbb{C}})^*$.

THEOREM 3.7.1. *The $(\mathbb{C}, A(N))$-Lie subalgebra P of $\Omega^1(N, \mathbb{C})_{\{\cdot,\cdot\}}$ is a complex analytic stratified Kähler polarization for the (fine) stratified symplectic space $(N, C^\infty(N), \{\cdot, \cdot\})$; the underlying complex analytic structure is that induced from the embedding of N into $\mathfrak{p}(\cong \mathfrak{p}^+)$.*

PROOF. Since each stratum of N inherits a smooth complex structure from the complex analytic structure of N, the symplectic stratification is a refinement of the complex analytic one. Let $w_1, \dots, w_\ell$ be a (complex) basis of $\mathfrak{p}^+$ and view the w_j's as holomorphic coordinate functions on $(\mathfrak{p}^+)^*$; restricting them to $\mathcal{O}$ yields holomorphic functions on $\mathcal{O}$, and every holomorphic function in $\mathcal{O}$ may be written as a holomorphic function in these coordinates. This implies that, given two holomorphic functions f and h in $\mathcal{O}$, the Poisson bracket $\{f, h\}$ vanishes, since the Lie bracket on $\mathfrak{p}^+$ (which yields the brackets of the kind $\{w_j, w_k\}$) is zero. Consequently the sheaf of germs of holomorphic functions on N is contained in the sheaf of germs of polarized functions. In view of Theorem 2.5, P is a stratified Kähler polarization which is compatible with the complex analytic structure; this may also be seen directly.

To see that the sheaf of germs of holomorphic functions coincides with that of polarized ones, we argue as follows: The top stratum $\mathcal{O}_r$ (i. e. the principal holomorphic nilpotent orbit) is Zariski open in $\mathfrak{p}^+$ and has $\mathfrak{p}^+$ as its closure. The complement $\overline{\mathcal{O}_{r-1}}$ of $\mathcal{O}_r$ in $\mathfrak{p}^+$ is a thin set in $\mathfrak{p}^+$. Hence a continuous function in $\mathfrak{p}^+$ which is holomorphic in $\mathcal{O}_r$ is holomorphic in $\mathfrak{p}^+$ **[17]**. This implies the assertion, since the complex analytic structure on the closure of any smaller holomorphic nilpotent orbit is induced from the embedding thereof into $\mathfrak{p}^+$. □

REMARK 3.7.2. While $\mathfrak{p}$ carries a symplectic structure as well which combines with its complex structure J_z to a Kähler structure, the symplectic structures on the strata of N *cannot* be induced from a symplectic structure on $\mathfrak{p}$, though.

REMARK 3.7.3. The image of $A(N) \otimes_{\mathbb{C}} \mathfrak{p}^-$ in $\Omega^1(N, \mathbb{C})$ is a stratified complex polarization $\overline{P} \subseteq \Omega^1(N, \mathbb{C})$ as well which is generated by the differentials dx where $x \in \mathfrak{p}^-$; it is just the complex conjugate of P. The $C^\infty(N, \mathbb{C})$-submodule of $\Omega^1(N, \mathbb{C})$ generated by P and $\overline{P}$ decomposes into a direct sum $P \oplus \overline{P}$. However, $\Omega^1(N, \mathbb{C})$ does *not* exhaust this direct sum. In fact, the inclusion of $P \oplus \overline{P}$ into $\Omega^1(N^{\mathrm{red}}, \mathbb{C})$ gives rise to an exact sequence

$$0 \longrightarrow P \oplus \overline{P} \longrightarrow \Omega^1(N, \mathbb{C}) \longrightarrow C \longrightarrow 0 \tag{3.7.4}$$

of $C^\infty(N, \mathbb{C})$-modules, the cokernel C being generated by (classes of) the differentials dw where $w \in \mathfrak{k}^{\mathbb{C}}$. The precise information about holomorphic nilpotent orbits spelled out in Theorems 3.5.4, 3.5.5, and 3.6.2 above yields at once explicit examples where C is non-zero. Thus a complex polarization can *no longer be thought of as being given by the* $(0,1)$*-vectors of a complex structure.* On any stratum, the constituent C is zero, though, and the module of differentials does of course decompose into the direct sum of the holomorphic and antiholomorphic polarization: Restricted to a stratum S, the exact sequence (3.7.4) comes down to an isomorphism $P_S \oplus \overline{P}_S \longrightarrow \Omega^1(S, \mathbb{C})$.

3.7.5. DEPENDENCE ON THE CHOICE OF H-ELEMENT. Given a nilpotent orbit $\mathcal{O}$, we have observed above (Addendum to Theorem 3.3.3) that the property of being holomorphic does not depend on the choice of H-element. The induced complex structure on the closure $\overline{\mathcal{O}}$ depends on that choice, though. Letting the H-element z vary, we obtain a moduli space of complex structures compatible with the other structure, i. e. a moduli space of complex analytic stratified Kähler structures on $\overline{\mathcal{O}}$, the stratified symplectic structure being fixed. This moduli space amounts to the homogeneous space G/K.

3.7.6. THE HOLOMORPHIC SEMISIMPLE ORBITS. In the same fashion, for any semisimple holomorphic orbit $\mathcal{O}$, a choice of H-element z induces a smooth K-invariant complex structure on $\mathcal{O} \cong G/K$ turning the latter into a Kähler manifold. This is *not* the standard hermitian symmetric space structure on G/K, though, since only K acts by isometries, and the curvature tensor is not parallel. A special case has been illustrated in (3.2.2) above.

4. Reduction and stratified Kähler spaces

In this section we will show that reduction is another source for stratified Kähler spaces. For later reference, it will be convenient to extend the usual reduction scheme in symplectic geometry.

Let $(N, C^\infty(N))$ be a stratified space, with smooth structure $C^\infty(N)$, and let G be a Lie group, with Lie algebra $\mathfrak{g}$. We will say that an action of G on N is *smooth* provided it preserves strata and carries smooth functions on N to smooth functions on N, that is, (i) given $h \in C^\infty(N)$ and $y \in G$, the function $f \circ y$ (where we do not distinguish in notation between $y \in G$ and the homeomorphism it induces on N) is in $C^\infty(N)$ and (ii), given $h \in C^\infty(N)$ and $Y \in \mathfrak{g}$, the function $Y_N h$ is in $C^\infty(N)$; here Y_N refers to the derivation which, on each stratum, is given by the vector field Y_N induced from Y on that stratum. As a smooth function, $Y_N h$ initially exists only stratumwise, whence the requirement (ii). We refer to a derivation of the kind Y_N as the *stratified vector field* on N induced from Y.

Let $\{\ ,\ \}$ be a Poisson structure on $C^\infty(N)$ which, on each stratum of N, restricts to a smooth (not necessarily symplectic) Poisson structure. We will refer to $(N, C^\infty(N), \{\ ,\ \})$ as a *stratified Poisson space*. We will say that the G-action on N is *hamiltonian*, with *stratified Poisson momentum mapping* $\mu \colon N \to \mathfrak{g}^*$, provided (i) μ is a smooth G-equivariant map, and (ii) for every $Y \in \mathfrak{g}$ and every smooth function f on N,

$$\{f, \mu^Y\} = Y_N f,$$

where $\mu^Y \colon N \to \mathbb{R}$ is the composite of μ with Y, viewed as a linear map on $\mathfrak{g}^*$, and where Y_N refers to the stratified vector field on N induced by the G-action. Here "smooth" means that, for every ordinary smooth function h on $\mathfrak{g}^*$, the composite $h \circ \mu$ is in $C^\infty(N)$. The requirement (ii) is equivalent to the adjoint $\delta \colon \mathfrak{g} \to C^\infty(N)$, which is given by $\delta(Y) = \mu^Y$ $(Y \in \mathfrak{g})$, having the property that (ii′) for every $Y \in \mathfrak{g}$, the hamiltonian vector field $\{\cdot, \mu^Y\}$ of $\delta(Y)$ coincide with the vector field Y_N; the mapping δ is then occasionally referred to as *comomentum*. Furthermore, when (ii) holds, the G-equivariance of μ implies that (i′) $\delta[Y,Z] = \{\delta(Y), \delta(Z)\}$, for every $Y, Z \in \mathfrak{g}$ and, for connected G, this is actually equivalent to the G-invariance. Notice that (i′) says that μ is a Poisson map where $\mathfrak{g}^*$ is endowed with its Lie-Poisson structure. When N is a smooth symplectic manifold or, more generally, a smooth Poisson manifold, this notion of Poisson momentum mapping amounts to the ordinary one in symplectic geometry or, more generally, to that of Poisson momentum mapping [**63**]. When $(N, C^\infty(N), \{\ ,\ \})$ is a stratified symplectic space acted upon by a Lie group G in a hamiltonian fashion with stratified Poisson momentum mapping $\mu \colon N \to \mathfrak{g}^*$, we will refer to μ as a *stratified symplectic space momentum mapping*. Examples which are not ordinary momentum mappings will be given in Section 8 below.

Let $(N, C^\infty(N), \{\ ,\ \})$ be a stratified Poisson space, and let G be a Lie group which acts smoothly on N in a hamiltonian fashion with stratified Poisson momentum mapping $\mu \colon N \to \mathfrak{g}^*$. Generalizing the standard construction, define the *reduced space* N^{red} by $N^{\mathrm{red}} = \mu^{-1}(0)/G$. The construction of the ARMS-CUSHMAN-GOTAY-algebra [**1**] extends as well: Define

$$C^\infty(N^{\mathrm{red}}) \quad \text{as} \quad \left(C^\infty(N)^G\right)/I^G,$$

the algebra $C^\infty(N)^G$ of smooth G-invariant functions on N, modulo the ideal I^G of functions in $C^\infty(N)^G$ that vanish on the zero locus $\mu^{-1}(0)$, so that $C^\infty(N^{\mathrm{red}})$

yields a *smooth structure* $C^\infty(N^{\mathrm{red}})$, i. e. algebra of continuous functions on N^{red} in an obvious fashion.

PROPOSITION 4.1. *The Poisson structure on $C^\infty(N)$ induces a Poisson structure $\{\cdot,\cdot\}^{\mathrm{red}}$ on $C^\infty(N^{\mathrm{red}})$.*

PROOF. Let f be a G-invariant function in $C^\infty(N)$. Noether's theorem is still true and says that μ is constant along the trajectories of $X_f = \{\cdot, f\}$; indeed, any trajectory lies in a stratum. Hence, given a function h in $C^\infty(N)$ which vanishes on $\mu^{-1}(0)$, for any $q \in \mu^{-1}(0)$, since $\mu(\exp(tX_f)q) = \mu(q)$, we have

$$\{h, f\}(q) = X_f(h)|_q = \frac{d}{dt}(h(\exp(tX_f)q))|_{t=0} = 0,$$

that is, $\{h, f\}$ vanishes on $\mu^{-1}(0)$ as well. Consequently I^G is a Poisson ideal in $C^\infty(N)^G$. □

We now suppose that G is compact. Given a symplectic manifold N acted upon by G in a hamiltonian fashion, each connected component of the reduced space N^{red} is a stratified symplectic space [**85**]; to simplify the exposition somewhat we shall not distinguish in notation between the reduced space and any one of its connected components. The existence of G-invariant partitions of unity on N with respect to arbitrary locally finite coverings by G-invariant open subsets entails that $C^\infty(N^{\mathrm{red}})$ is fine.

PROPOSITION 4.2. *Given a positive Kähler manifold N with a holomorphic $G^{\mathbb{C}}$-action whose restriction to (a compact real form) G preserves the Kähler structure and is hamiltonian, the Kähler polarization F induces a (positive) normal Kähler structure on the reduced space N^{red}.*

PROOF. This follows readily from the results in [**23**] and [**84**], combined with Theorem 2.5 above. Indeed, the zero locus $\mu^{-1}(0)$ is a *Kempf-Ness* set of N, and the inclusion of $\mu^{-1}(0)$ into N induces a homeomorphism

$$N^{\mathrm{red}} = \mu^{-1}(0)/G \to N//G^{\mathbb{C}}$$

from the reduced space onto the G(eometric) I(nvariant) T(heory) quotient $N//G^{\mathbb{C}}$. The latter inherits a complex analytic structure which, locally, amounts to that of a Stein quotient in such a way that the symplectic stratification of N^{red} is a refinement of the complex analytic one and that each stratum, with its induced complex analytic structure, inherits a Kähler metric; see (2.3) of [**84**] (under the additional technical hypothesis that the momentum mapping be "admissible") and [**23**]. For every pair f and h of holomorphic functions in N, the Poisson bracket $\{f, h\}$ manifestly vanishes, and this property descends to N^{red}. By Theorem 2.5, the $C^\infty(N^{\mathrm{red}}, \mathbb{C})$-submodule P of $\Omega^1(N^{\mathrm{red}}, \mathbb{C})_{\{\cdot,\cdot\}}$ described there (where $X = N^{\mathrm{red}}$) is a complex analytic stratified Kähler polarization. Normality is a consequence of the fact that, locally, the GIT-quotient may be written as a quotient of a smooth Stein space and, a smooth complex manifold being necessarily normal, passing to the quotient preserves normality. Since, complex analytically, the space N^{red} is normal, we might as well first observe that P is a stratified Kähler polarization on N^{red} such that the complex analytic structure is compatible with it and then deduce, using Proposition 2.4, that P is actually a normal Kähler structure. The positivity of P is obvious. □

EXAMPLE 4.3. Let Σ be a closed Riemann surface, let G be a compact Lie group, and let $\xi: P \to \Sigma$ be a principal G-bundle. Pick a Riemannian metric on Σ and an invariant positive definite symmetric bilinear form on the Lie algebra $\mathfrak{g}$ of G, and let N be the moduli space of central Yang-Mills connections on ξ determined by these data [**3**]. To recall its well known representation theory interpretation, write Z for the center of G, let $\mathfrak{z} = \mathrm{Lie}(Z)$, and pick $X \in \mathfrak{z}$. Let π be the fundamental group of Σ, let $0 \to \mathbb{Z} \to \Gamma \to \pi \to 1$ be its universal central extension, let $\mathrm{Hom}_X(\Gamma, G)$ be the space of homomorphisms from Γ to G which send a preferred generator of the kernel $\mathbb{Z}$ to $\exp X \in G$, and consider the space $\mathrm{Rep}_X(\Gamma, G) = \mathrm{Hom}_X(\Gamma, G)/G$ of G-orbits (under conjugation); depending on the choice of X, the space $\mathrm{Hom}_X(\Gamma, G)$ may be empty. The topological type of ξ determines an element $X \in \mathfrak{z}$ such that taking holonomies identifies the space N with a connected component of $\mathrm{Rep}_X(\Gamma, G)$ [**3**], [**33–36**]. We will refer to any of the connected components of $\mathrm{Rep}_X(\Gamma, G)$ as a *space of twisted representations* of π in G. As final ingredient, we pick a complex structure on Σ.

THEOREM 4.3.1. *The data induce on N a normal (complex analytic positive) Kähler structure turning N into an (in general) exotic projective variety.*

Explicit constructions of the stratified symplectic structure (which does not depend on the complex structure on Σ) may be found in [**33–36**]. For $G = \mathrm{U}(n)$, the requisite complex analytic structure on N was constructed by Narasimhan and Seshadri [**69**] by means of geometric invariant theory; the space N may then be identified with the moduli space of semistable holomorphic rank n vector bundles on Σ of a certain degree k which corresponds to the topological type of ξ or, equivalently, to the twisting of the representations of π; for $k = 0$ the representations of π in G are ordinary ones, i. e. are untwisted. For general G, a construction of the complex analytic structure may be found e. g. in [**58**].

It is not obvious that the two structures combine to a complex analytic Kähler structure, though, except in the (complex analytically or algebraically) non-singular case where the space N is known to be a Kähler manifold [**69**]. We now explain the requisite analytical details.

Let A be an arbitrary connection (not necessarily Yang-Mills) on ξ, and let d_A be its operator of covariant derivative. It is well known that A determines (and is determined by) a complex structure on the complexification $\xi^{\mathbb{C}}: P^{\mathbb{C}} \to \Sigma$ of ξ which turns $\xi^{\mathbb{C}}$ into a holomorphic principal $G^{\mathbb{C}}$-bundle; we write the latter as $\xi^{\mathbb{C}}_A$. The chosen invariant positive definite symmetric bilinear form on $\mathfrak{g}$ induces a hermitian structure on $\mathrm{ad}(\xi^{\mathbb{C}})$ which, in turn, determines a Hodge decomposition

$$\Omega^{0,0}(\Sigma, \mathrm{ad}(\xi^{\mathbb{C}})) = \mathcal{H}^{0,0}_A \oplus L^{0,0}, \quad \Omega^{0,1}(\Sigma, \mathrm{ad}(\xi^{\mathbb{C}})) = \mathcal{H}^{0,1}_A \oplus d''_A(L^{0,0}),$$

the spaces of harmonic forms being denoted by $\mathcal{H}^{0,*}_A$ and the Cauchy-Riemann operator by d''_A. A standard construction, cf. Lemma 2.1 (ii) in [**69**] for $G = \mathrm{U}(n)$, yields a complete analytic family of holomorphic principal $G^{\mathbb{C}}$-bundles $\{\xi_y\}_{y \in Y_A}$ parametrized by a based ball (Y_A, o) in the complex vector space $\mathcal{H}^{0,1}_A$ so that ξ_o "coincides" with $\xi^{\mathbb{C}}_A$, that is, a holomorphic principal $G^{\mathbb{C}}$-bundle over $Y_A \times \Sigma$ whose Kodaira-Spencer map $\mathrm{T}_o Y_A \to \mathcal{H}^{0,1}_A$ is an isomorphism, together with an isomorphism of holomorphic principal bundles between ξ_o and $\xi^{\mathbb{C}}_A$.

Suppose that A is a central Yang-Mills connection, and consider the real differential graded Lie algebra $(\Omega^*(\Sigma, \mathrm{ad}(\xi)), d_A)$. It controls the infinitesimal central

Yang-Mills connections variations of (ξ, A). The star operator $*: \Omega^*(\Sigma, \mathrm{ad}(\xi)) \to \Omega^{2-*}(\Sigma, \mathrm{ad}(\xi))$ determined by the data endows $\Omega^1(\Sigma, \mathrm{ad}(\xi))$ with a complex structure, and associating $\eta + i * \eta$ to η identifies $\Omega^1(\Sigma, \mathrm{ad}(\xi))$ with $\Omega^{0,1}(\Sigma, \mathrm{ad}(\xi^{\mathbb{C}}))$ as a complex vector space. Under this isomorphism, the space $\mathcal{H}^1_A$ of real harmonic 1-forms gets identified with $\mathcal{H}^{0,1}_A$. The space of connections $\mathcal{A}(\xi)$ being endowed with the standard symplectic structure ω given by $\omega(\alpha, \beta) = \int_\Sigma \langle \alpha \wedge \beta \rangle$ where $\alpha, \beta \in \Omega^1(\Sigma, \mathrm{ad}(\xi))$ and where $\langle \, , \rangle$ is the pairing induced from the given invariant symmetric bilinear form on $\mathfrak{g}$, the assignment to a connection of its curvature is a momentum mapping for the action of the group of gauge transformations. Consider the symplectic slice $\mathcal{M}_A \subseteq \mathcal{A}(\xi)$ given in (2.16) of [**31**]. The symplectic structure ω induces a symplectic structure ω_A on $\mathcal{M}_A$. Write Z_A for the stabilizer of A in the group of gauge transformations of ξ, and let $\mathfrak{z}_A = \mathrm{Lie}(Z_A)$. The momentum mapping for the action of the group of gauge transformations restricts to a momentum mapping $\mu_A: \mathcal{M}_A \to \mathcal{H}^2_A$ where $\mathcal{H}^2_A \cong \mathrm{H}^2(\Sigma, \mathrm{ad}(\xi))$ is identified with the dual of $\mathfrak{z}_A$ in the standard fashion. After having $\mathcal{M}_A$ suitably cut to size (if need be), when we associate to a connection $A + \eta$ in $\mathcal{M}_A$ its induced Cauchy-Riemann operator $d''_{A+\eta}$, we obtain a map from $\mathcal{M}_A$ to the GIT quotient $Y_A//Z^{\mathbb{C}}_A$ which induces a homeomorphism from the reduced space $\mu_A^{-1}(F_\xi)/Z_A$ onto $Y_A//Z^{\mathbb{C}}_A$; here F_ξ denotes the corresponding constant central curvature. On the other hand, an explicit map from $\mu_A^{-1}(F_\xi)/Z_A$ to N, realized as a connected component of a space of the kind $\mathrm{Rep}_X(\Gamma, G)$, is given by the assignment to a point $A + \eta$ of its holonomies with respect to a system of curves representing the generators of the fundamental group. See [**38**] for details. Letting A vary (among central Yang-Mills connections), we obtain these structures on suitable "coordinate patches" of the space N. The Poisson structure comes from the reduced spaces of the kind $\mu_A^{-1}(F_\xi)/Z_A$, while the complex analytic one is inherited from the complex GIT quotient's $Y_A//Z^{\mathbb{C}}_A$. The global stratified symplectic Poisson structure on $\mathrm{Rep}_X(\Gamma, G)$ is that constructed in [**35**]. Thus we obtain a normal Kähler structure on N, the normality being an immediate consequence of the local complex GIT quotient constructions. More details will be given at another occasion.

Apart from the Hodge decomposition of $\Omega(\Sigma, \mathrm{ad}(\xi^{\mathbb{C}}))$ on which the existence of the complete families of holomorphic principal $G^{\mathbb{C}}$-bundles relies and the Banach space techniques which yield the slice $\mathcal{M}_A$, *this construction of the moduli space N as a normal Kähler space is finite dimensional and entirely analytic, that is, it avoids the usual detour via geometric invariant theory.* Normality is a consequence of the fact that, locally, the space is an affine geometric invariant theory quotient. The *complex analytic stratified Kähler structure* isolated here is somewhat finer than the complex analytic structure alone, though, and cannot be obtained from geometric invariant theory. *This stratified Kähler structure gives a precise description of the Kähler singularity behaviour.* The issue of singularities was raised in [**3**]. When $G = \mathrm{SU}(2)$, as already pointed out in the introduction, the space N is an exotic $\mathbb{C}\mathrm{P}^3$, and the complement of the top stratum is a Kummer surface which, in turn, is the singular locus, in the sense of complex analytic stratified Kähler spaces, of this exotic $\mathbb{C}\mathrm{P}^3$. For various special cases, the local structure of the space N near any of its points has been examined in [**33**]; see also [**39–41**].

More generally, certain moduli spaces including moduli spaces of semistable holomorphic vector bundles with parabolic structure [**66**] inherit normal Kähler structures, the requisite stratified symplectic structure being that constructed in [**20**]. We will explain the details elsewhere.

5. Associated representations and singular reduction

In this section we will describe the holomorphic nilpotent orbits for the classical cases as reduced spaces for suitable momentum mappings.

5.1. INVARIANTS AND MOMENTUM MAPPING. Consider a real finite-dimensional symplectic vector space W, with symplectic form ω. The algebra $\mathbb{R}[W^*]$ of real polynomial functions on W—after a choice of basis $b_1, \ldots, b_n$: the algebra of polynomials in the coordinate functions $x_1, \ldots, x_n$—inherits a Poisson structure $\{\cdot,\cdot\}$ from the symplectic structure which is given by $\{x_j, x_k\} = \omega(b_j, b_k)$, $1 \leq j, k \leq n$; furthermore, the homogeneous quadratic part $\mathbb{R}_2[W^*]$ of $\mathbb{R}[W^*]$ is closed under the (Poisson) bracket and, in fact, as a Lie algebra, isomorphic to $\mathfrak{sp}(W,\omega)$. An explicit isomorphism $\delta\colon \mathfrak{sp}(W,\omega) \to \mathbb{R}_2[W^*]$ is given by the assignment to $X \in \mathfrak{sp}(W,\omega)$ of the quadratic polynomial function f_X on W where $f_X(v) = \frac{1}{2}\omega(Xv, v)$ $(v \in W)$. The inverse map sends a quadratic polynomial f to its Hamiltonian vector field $X_f = \{\cdot, f\}$. More precisely, the assignment to $X \in \mathfrak{sp}(W,\omega)$ of the Hamiltonian vector field of f_X is the linear vector field on W induced by X, the corresponding map from $\mathfrak{sp}(W,\omega)$ to $\mathrm{Vect}(W)$ being an *anti* Lie isomorphism onto the Lie algebra of linear vector fields. For $W = \mathbb{R}^2$, with coordinates q, p and ω the standard form so that $\{q,p\} = 1$, with the notation E, F, H in (3.2.2),

$$\delta(E) = \frac{p^2}{2}, \quad \delta(F) = -\frac{q^2}{2}, \quad \delta(H) = pq.$$

For general W and ω, the unique momentum mapping $\mu\colon W \to (\mathbb{R}_2[W^*])^*$ (having the value zero at the origin) for the $\mathrm{Sp}(W,\omega)$-action on W is just the adjoint of δ and therefore simply given by the formula

$$(f \circ \mu)(v) = f(v), \quad v \in W,\ f \in \mathbb{R}_2[W^*],$$

which, in terms of $\mathfrak{sp}(W,\omega)$, comes down to the well known formula

$$(X \circ \mu)(v) = \frac{1}{2}\omega(Xv, v), \quad v \in W,\ X \in \mathfrak{sp}(W,\omega).$$

Under the canonical isomorphism between $(\mathbb{R}_2[W^*])^*$ and the symmetric square $\mathrm{S}^2_{\mathbb{R}}[W]$ of W, this momentum mapping amounts to the canonical squaring map for W. Given $v \in W$, define $v^\dagger\colon W \to \mathbb{R}$ by $\omega(u, v) = v^\dagger u$; the composite $vv^\dagger$, viewed as a linear endomorphism of W, then lies in $\mathfrak{sp}(W,\omega)$. Plainly, for $v \in W$,

$$(5.1.1) \qquad (X \circ \mu)(v) = \frac{1}{2}v^\dagger X v = \frac{1}{2}\mathrm{trace}(X \circ (vv^\dagger)),$$

that is to say, when we identify $\mathfrak{sp}(W,\omega)$ with its dual by means of the $\mathrm{Sp}(W,\omega)$-invariant pairing $(A, B) \mapsto \frac{1}{2}\mathrm{trace}(AB)$, referred to henceforth as the *half-trace pairing*—it is a positive multiple of the Killing form—the momentum mapping μ amounts to the assignment to $v \in W$ of $vv^\dagger \in \mathfrak{sp}(W,\omega)$.

EXAMPLE. Let $W = \mathbb{R}^4$, with standard symplectic structure which, in the coordinates q^1, q^2, p^1, p^2, amounts to the Poisson brackets $\{q^j, p^j\} = 1$ $(j = 1, 2)$ etc. With the notation $v = [q^1, q^2, p^1, p^2]^\tau$, we obtain

$$\mu(v) = vv^\dagger = \begin{bmatrix} q^1p^1 & q^1p^2 & -q^1q^1 & -q^1q^2 \\ q^2p^1 & q^2p^2 & -q^2q^1 & -q^2q^2 \\ p^1p^1 & p^1p^2 & -p^1q^1 & -p^1q^2 \\ p^2p^1 & p^2p^2 & -p^2q^1 & -p^2q^2 \end{bmatrix}.$$

We may identify W with the total space $\mathrm{T}^*(\mathbb{R}^2)$ of the cotangent bundle of the ordinary (q^1, q^2)-plane $\mathbb{R}^2$. Ordinary angular momentum in the plane has the infinitesimal generator $X = \begin{bmatrix} 0 & -1 & 0 & 0 \\ 1 & 0 & 0 & 0 \\ 0 & 0 & 0 & -1 \\ 0 & 0 & 1 & 0 \end{bmatrix}$, and

$$(X \circ \mu)(v) = \frac{1}{2}\mathrm{trace}(X \circ (vv^\dagger)) = q^1p^2 - q^2p^1.$$

This is the angular momentum mapping for a single particle in the plane.

For general W and ω, given a subgroup H of $\mathrm{Sp}(W, \omega)$, the induced H-action on W induces an H-action on $\mathbb{R}_2[W^*]$ preserving the Poisson structure, and the invariants $\mathfrak{g} = \mathbb{R}_2[W^*]^H$ constitute a Lie subalgebra of $\mathbb{R}_2[W^*]$. Indeed, in $\mathfrak{sp}(W)$, $\mathfrak{g}$ amounts to the H-invariant subspace of $\mathfrak{sp}(W)$, with reference to the adjoint representation. The H-invariant map $W \to \mathfrak{g}^*$ whose adjoint is the injection $\mathfrak{g} \to \mathbb{R}[W^*]$ is plainly a momentum mapping for the induced action of the Lie group $G \subseteq \mathrm{Sp}(W, \omega)$ with $\mathrm{Lie}(G) = \mathfrak{g}$. For example, when H is the obvious copy of $\mathbb{Z}/2$ in $\mathrm{Sp}(W, \omega)$ whose generator acts on W via multiplication by -1, $\mathbb{R}_2[W^*]^H = \mathbb{R}_2[W^*]$, the homogeneous degree 2 part $\mathbb{R}_2[W^*]$ generates the algebra of $\mathbb{Z}/2$-invariants, and the momentum mapping μ amounts to the Hilbert map of invariant theory for the $\mathbb{Z}/2$-action on W.

5.2. Associated representations. Given a unitary representation E of a compact Lie group H, with $\mathfrak{h} = \mathrm{Lie}(H)$, the construction in (5.1) applies to $W = E$ endowed with the symplectic structure coming from the hermitian form on E and yields the unique momentum mapping $\mu_H\colon E \to \mathfrak{h}^*$ having the value zero at the origin of E. Given a nilpotent orbit $\mathcal{O}$ in a real semisimple Lie algebra $\mathfrak{g}$, we will say that a unitary representation E of a compact Lie group H is a *compact associated representation* for $\mathcal{O}$ provided $\mathbb{R}_2[E^*]^H \cong \mathfrak{g}$ in such a way that the H-invariant map $\mu_G\colon E \to \mathfrak{g}^*$ whose adjoint is the injection $\mathfrak{g} \to \mathbb{R}[E^*]$ induces an isomorphism of stratified symplectic spaces from the H-reduced space $E^{\mathrm{red}} = \mu_H^{-1}(0)/H$ onto the closure $\overline{\mathcal{O}}$ of $\mathcal{O}$ in $\mathfrak{g}^*$.

More generally, let H be an arbitrary Lie group, not necessarily compact, with Lie algebra $\mathfrak{h}$, which acts linearly and symplectically on a symplectic vector space E, with momentum mapping $\mu_H\colon E \to \mathfrak{h}^*$ having the value zero at the origin of E. We then refer to E as a *symplectic representation* of H and *define* the reduced space E^{red} to be the space of *closed* orbits in the zero locus $\mu_H^{-1}(0)$. Given a nilpotent orbit $\mathcal{O}$ in a real semisimple Lie algebra $\mathfrak{g}$, we will say that a symplectic representation E of a Lie group H is an *associated representation* for $\mathcal{O}$ provided $\mathbb{R}_2[E^*]^H \cong \mathfrak{g}$ in such a way that the H-invariant map $\mu_G\colon E \to \mathfrak{g}^*$ whose adjoint is the injection $\mathfrak{g} \to \mathbb{R}[E^*]$ induces a Poisson map $\overline{\mu}_G\colon E^{\mathrm{red}} \to \mathfrak{g}^*$ which is an isomorphism of stratified spaces from E^{red} onto the closure $\overline{\mathcal{O}}$ of $\mathcal{O}$ in $\mathfrak{g}^*$.

5.3. Compact associated representations. We will now establish their existence for *holomorphic* nilpotent orbits in the *standard cases*. Our tool will be an extension of the *basic construction* in [**55**]; in that reference, this construction is given only for the case of the classical groups over the complex numbers. We now extend this construction to the appropriate real forms of the classical groups. A key step will be an extension of the "First Main Theorem of Invariant Theory" to the real forms of the classical groups, given as Theorem 5.3.1 below. The significance of the First Main Theorem of Invariant Theory for orbits of complex classical groups has been noticed in [**55**].

Let $\mathbb{K} = \mathbb{R}, \mathbb{C}, \mathbb{H}$, let $V^s = \mathbb{K}^s$, the standard (right) $\mathbb{K}$-vector space of dimension s, endowed with a non-degenerate hermitian form $(\cdot,\cdot)$ of type (s', s''), that is, of rank $s = s' + s''$ and signature $s' - s''$; further, let $V = \mathbb{K}^n$, endowed with a complex structure J_V and skew form $\mathcal{B}$ of the kind reproduced in (3.5), let $H = H(s', s'') = \mathrm{U}(V^s, (\cdot,\cdot))$, $\mathfrak{h} = \mathrm{Lie}(H)$, $G = \mathrm{U}(V, \mathcal{B})$, and $\mathfrak{g} = \mathrm{Lie}(G)$ (cf. (3.5)), and denote the split rank of $G = \mathrm{U}(V, \mathcal{B})$ by r. When $s = s'$ and $s'' = 0$ we will write $H(s)$ instead of $H(s', s'')$. More explicitly:
(1) $\mathbb{K} = \mathbb{R}$, $V^s = \mathbb{R}^s$, $V = \mathbb{R}^n$, $n = 2\ell$, $H = \mathrm{O}(s', s'')$, $G = \mathrm{Sp}(\ell, \mathbb{R})$, $r = \ell$;
(2) $\mathbb{K} = \mathbb{C}$, $V^s = \mathbb{C}^s$, $V = \mathbb{C}^n$, $n = p + q$, $H = \mathrm{U}(s', s'')$, $G = \mathrm{U}(p, q)$, $p \geq q$, $r = q$;
(3) $\mathbb{K} = \mathbb{H}$, $V^s = \mathbb{H}^s$, $V = \mathbb{H}^n$, $H = \mathrm{U}(s', s'', \mathbb{H}) = \mathrm{Sp}(s', s'')$, $G = \mathrm{O}^*(2n)$, $r = [\frac{n}{2}]$.
When $s = s'$ and $s'' = 0$, $H = H(s)$ is the group $\mathrm{O}(s, \mathbb{R})$, $\mathrm{U}(s)$, or $\mathrm{Sp}(s)$ as appropriate.

Let $W = W(s) = \mathrm{Hom}_{\mathbb{K}}(V^s, V)$. The groups H and G act on W in an obvious fashion: Given $x \in H$, $\alpha \in \mathrm{Hom}_{\mathbb{K}}(V^s, V)$, $y \in G$, the action is given by the assignment to (x, y, α) of $y\alpha x^{-1}$. Given a $\mathbb{K}$-linear map $\alpha\colon V^s \to V$, define the $\mathbb{K}$-linear map $\alpha^\dagger\colon V \to V^s$ by

$$(\alpha^\dagger \mathbf{u}, \mathbf{v}) = \mathcal{B}(\mathbf{u}, \alpha\mathbf{v}), \ \mathbf{u} \in V, \ \mathbf{v} \in V^s;$$

given $x \in H, y \in G$, plainly $(y\alpha x^{-1})^\dagger = x\alpha^\dagger y^{-1}$. In case (1), for $s = 1$, this construction comes down to the assignment to $v \in W$ of $v^\dagger\colon W \to \mathbb{R}$ given in (5.1). For intelligibility we give explicit formulas for $\alpha^\dagger$ for the special cases where $s' = s$ (arbitrary value) and $s'' = 0$.
Case (1): $\alpha = [\mathbf{q}_1, \dots, \mathbf{q}_\ell, \mathbf{p}_1, \dots, \mathbf{p}_\ell]^\tau$:

$$\alpha^\dagger = [\mathbf{p}_1, \dots, \mathbf{p}_\ell, -\mathbf{q}_1, \dots, -\mathbf{q}_\ell]\colon \mathbb{R}^n \to \mathbb{R}^s,$$

$\mathbf{q}_1, \dots, \mathbf{q}_\ell, \mathbf{p}_1, \dots, \mathbf{p}_\ell \in \mathbb{R}^s$ being the column vectors of $\alpha^\dagger$.
Case (2): $\alpha = [\mathbf{w}_1, \dots, \mathbf{w}_{p+q}]^\tau$:

$$\alpha^\dagger = i[\overline{\mathbf{w}_1}, \dots, \overline{\mathbf{w}_p}, -\overline{\mathbf{w}_{p+1}}, \dots, -\overline{\mathbf{w}_{p+q}}]\colon \mathbb{C}^{p+q} \to \mathbb{C}^s,$$

$\mathbf{w}_1, \dots, \mathbf{w}_{p+q} \in \mathbb{C}^s$ being the column vectors of $\alpha^\dagger$.
Case (3): $\alpha = [\mathbf{w}_1, \dots, \mathbf{w}_n]^\tau$:

$$\alpha^\dagger = [\overline{\mathbf{w}_1}\mathcal{J}, \dots, \overline{\mathbf{w}_n}\mathcal{J}]\colon \mathbb{H}^n \to \mathbb{H}^s,$$

$\mathbf{w}_1, \dots, \mathbf{w}_n \in \mathbb{H}^s$ being the column vectors of $\alpha^\dagger$.

The 2-forms $(\cdot,\cdot)$ and $\mathcal{B}$ induce a symplectic structure ω_W on W by means of the assignment

$$\omega_W(\alpha, \beta) = \mathrm{trace}_{\mathrm{r}}(\beta^\dagger\alpha), \quad \alpha, \beta \in W,$$

where $\mathrm{trace}_{\mathrm{r}}(\beta^\dagger\alpha) \in \mathbb{R}$ is the *real* constituent of the value of the trace of $\beta^\dagger\alpha$. In case (1) where $\mathbb{K} = \mathbb{R}$ the function $\mathrm{trace}_{\mathrm{r}}$ is the ordinary trace function. In case (2) and case (3), the hermitian form $(\cdot,\cdot)$ on $V^s = \mathbb{K}^s$ determines a non-degenerate real symmetric bilinear form $(\cdot,\cdot)_r$ on $\mathbb{R}^s$ of the same signature as $(\cdot,\cdot)$ such that the *real* constituent of $(\cdot,\cdot)$ may be written as $(\mathbb{R}^s, (\cdot,\cdot)_r) \otimes_{\mathbb{R}} \mathbb{K}$, and the real constituent $\mathcal{B}_1$ of $\mathcal{B}$ is a real symplectic structure on V as observed earlier. Writing W in the form $W = \mathrm{Hom}_{\mathbb{R}}(\mathbb{R}^s, V)$, given $\alpha\colon \mathbb{R}^s \to V$, we may define $\alpha^{\dagger\dagger}\colon V \to \mathbb{R}^s$ by $(\alpha^{\dagger\dagger}\mathbf{u}, \mathbf{v})_r = \mathcal{B}_1(\mathbf{u}, \alpha\mathbf{v})$ where $\mathbf{u} \in V$ and $\mathbf{v} \in \mathbb{R}^s$; then

$$\omega_W(\alpha, \beta) = \mathrm{trace}(\beta^{\dagger\dagger}\alpha), \quad \alpha, \beta \in W,$$

where "trace" now refers to the ordinary real trace on $\mathrm{End}(\mathbb{R}^s)$, and where we do not distinguish in notation between an $\mathbb{R}$-linear map α from $\mathbb{R}^s$ to V and its $\mathbb{K}$-linear extension to V^s.

The actions of H and G on W preserve the symplectic structure ω_W and are hamiltonian. Up to the requisite identifications of $\mathfrak{h}$ and $\mathfrak{g}$ with their duals by means of the half-trace pairing, cf. (5.1) above, the momentum mappings μ_H and μ_G for the H- and G-actions are given by

$$\mu_H\colon W \longrightarrow \mathfrak{h}, \quad \mu_H(\alpha) = -\alpha^\dagger\alpha\colon V^s \to V^s,$$
$$\mu_G\colon W \longrightarrow \mathfrak{g}, \quad \mu_G(\alpha) = \alpha\alpha^\dagger\colon V \to V,$$

respectively. Indeed, since the H-action on W is given by the assignment to (x, α) of αx^{-1} ($x \in H$, $\alpha \in W$) the $\mathfrak{h}$-action on W is given by the formula

$$X(\alpha) = -\alpha X, \quad X \in \mathfrak{h}, \alpha \in W.$$

Hence, cf. (5.1.1),

$$(X \circ \mu_H)(\alpha) = \frac{1}{2}\omega_W(X(\alpha), \alpha) = -\frac{1}{2}\omega_W(\alpha X, \alpha) = -\frac{1}{2}\mathrm{trace}_{\mathrm{r}}((\alpha^\dagger\alpha) \circ X).$$

This shows that μ_H is indeed the H-momentum mapping. A similar reasoning justifies the claim that μ_G is the G-momentum mapping.

We could have identified $\mathfrak{h}$ and $\mathfrak{g}$ with their duals by means of the *trace* pairing, but we prefer *not* to do so since, up to sign, with our conventions, the momentum mappings will appear as the Hilbert maps of invariant theory.

ILLUSTRATION. Consider the case (1) with $s' = s$ and $s'' = 0$ so that $H = \mathrm{O}(s, \mathbb{R})$. Given $\mathbf{x} = (x^1, \dots, x^s) \in \mathbb{R}^s$ and $\mathbf{y} = (y^1, \dots, y^s) \in \mathbb{R}^s$, with the standard notation

$$\mathbf{x} \wedge \mathbf{y} \in \Lambda^2_{\mathbb{R}}[\mathbb{R}^s] = \mathfrak{so}(s, \mathbb{R})$$

for the skew-symmetric matrix having $x^j y^k - x^k y^j$ as its (j,k)'th entry, when the linear map $\alpha\colon V^s \to V$ is written as $\alpha = [\mathbf{q}_1, \dots, \mathbf{q}_\ell, \mathbf{p}_1, \dots, \mathbf{p}_\ell]^\tau$ with row vectors $\mathbf{q}_1, \dots, \mathbf{q}_\ell, \mathbf{p}_1, \dots, \mathbf{p}_\ell \in \mathbb{R}^s$, the values $\mu_H(\alpha) = -\alpha^\dagger\alpha \in \mathfrak{so}(s, \mathbb{R})$ and $\mu_G(\alpha) = \alpha\alpha^\dagger \in \mathfrak{sp}(\ell, \mathbb{R})$ admit the following descriptions:

$$\mu_H(\alpha) = -\alpha^\dagger\alpha = \mathbf{q}_1 \wedge \mathbf{p}_1 + \dots + \mathbf{q}_\ell \wedge \mathbf{p}_\ell \in \Lambda^2_{\mathbb{R}}[\mathbb{R}^s] = \mathfrak{so}(s, \mathbb{R})$$
$$\mu_G(\alpha) = \alpha\alpha^\dagger = \begin{bmatrix} [\mathbf{q}_j\mathbf{p}_k] & -[\mathbf{q}_j\mathbf{q}_k] \\ [\mathbf{p}_j\mathbf{p}_k] & -[\mathbf{p}_j\mathbf{q}_k] \end{bmatrix} \in \mathfrak{sp}(\ell, \mathbb{R}).$$

Here $\mathfrak{so}(s, \mathbb{R})$ (as well as $\mathfrak{sp}(\ell, \mathbb{R})$) is identified with its dual via the half-trace pairing $(A, B) \mapsto \frac{1}{2}\mathrm{trace}(AB)$. The map μ_H comes down to the ordinary angular momentum mapping for a constrained system of ℓ particles in $\mathbb{R}^s$ whence the notation $\mathbf{q}_1, \dots, \mathbf{q}_\ell, \mathbf{p}_1, \dots, \mathbf{p}_\ell$. For example, when $s = 2$ and $\ell = 1$, $W = \mathrm{End}_{\mathbb{R}}(\mathbb{R}^2)$ and, taking $\alpha = \begin{bmatrix} \mathbf{q} \\ \mathbf{p} \end{bmatrix} = \begin{bmatrix} q^1 & q^2 \\ p^1 & p^2 \end{bmatrix}$, we have $\alpha^\dagger = \begin{bmatrix} p^1 & -q^1 \\ p^2 & -q^2 \end{bmatrix}$ and $-\alpha^\dagger\alpha = \begin{bmatrix} 0 & q^1p^2 - p^1q^2 \\ q^2p^1 - p^2q^1 & 0 \end{bmatrix}$. When we identify W with the total space $\mathrm{T}^*(\mathbb{R}^2)$

of the cotangent bundle of the ordinary (q^1, q^2)-plane $\mathbb{R}^2$, taking the infinitesimal generator $X = \begin{bmatrix} 0 & -1 \\ 1 & 0 \end{bmatrix}$ of ordinary angular momentum, we have

$$(X \circ \mu_H)(\alpha) = -\frac{1}{2}\operatorname{trace}((\alpha^\dagger \alpha) \circ X) = q^1 p^2 - q^2 p^1.$$

This is again the angular momentum mapping for a single particle in the plane, the present description being a variant of the example in (5.1) above.

Let E be a (real or complex) symplectic vector space. Recall that two subgroups Γ and Γ' of $\mathrm{Sp}(E)$ are said to constitute a *reductive dual pair* provided Γ and Γ' are each other's full centralizer in $\mathrm{Sp}(E)$ [**27**]. In each of the cases (1),(2),(3) above, the two groups G and H constitute a real reductive dual pair in $\mathrm{Sp}(W)$ [**28**]; these are precisely the pairs (5.2)(i),(iii),(iv) in that reference. Below (G, H) refers to any of these three pairs.

THEOREM 5.3.1. *The ring of all H-invariant polynomials in $\mathbb{R}[W^*]$ is generated by its homogeneous quadratic part, and the canonical map from $\mathbb{R}_2[W^*]^H$ to $\mathfrak{g}$ is an isomorphism in such a way that the adjoint of the injection of $\mathfrak{g}$ into $\mathbb{R}[W^*]$ amounts to the G-momentum mapping $\mu_G\colon W \to \mathfrak{g}^*$. Likewise, the ring of all G-invariant polynomials in $\mathbb{R}[W^*]$ is generated by its homogeneous quadratic part, and the canonical map from $\mathbb{R}_2[W^*]^G$ to $\mathfrak{h}$ is an isomorphism such that the H-momentum mapping $\mu_H\colon W \to \mathfrak{h}^*$ is the adjoint of the injection of $\mathfrak{h}$ into $\mathbb{R}[W^*]$.*

COROLLARY 5.3.2. *The components of the μ_H-momentum mapping generate the algebra of G-invariants in $\mathbb{R}[W^*]$ and the components of the μ_G-momentum mapping generate the algebra of H-invariants in $\mathbb{R}[W^*]$.*

This Corollary extends a result in [**60**]. Its statement is a version of the "First Main Theorem of Invariant Theory" [**93**], cf. also [**27**]. However this theorem does *not* say that the H-invariants combine to a momentum mapping for G nor that the G-invariants combine to a momentum mapping for H. Moreover, this theorem is usually phrased only for the classical groups

$$\mathrm{O}(n,\mathbb{R}),\ \mathrm{O}(n,\mathbb{C}),\ \mathrm{Sp}(n,\mathbb{R}),\ \mathrm{Sp}(n,\mathbb{C}),\ \mathrm{GL}(n,\mathbb{R}),\ \mathrm{GL}(n,\mathbb{C}).$$

There is as well a close relationship between the situation of Theorem 5.3.1 and the notion of dual pair of Poisson mappings [**91**].

PROOF OF THEOREM 5.3.1. Since H and G constitute a reductive dual pair in $\mathrm{Sp}(W)$, the homogeneous degree 2 part $\mathbb{R}_2[W^*]^H$ coincides with (a copy of) $\mathfrak{g}$ and, likewise, the homogeneous degree 2 part $\mathbb{R}_2[W^*]^G$ coincides with (a copy of) $\mathfrak{h}$. We note that, by construction, $\mathfrak{g}$ and $\mathfrak{h}$ are thus realized within $\mathbb{R}_2[W^*] = \mathfrak{sp}(W)$, which equals $\mathfrak{sp}(s\ell,\mathbb{R})$ for $\mathbb{K} = \mathbb{R}$, $\mathfrak{sp}(sn,\mathbb{R})$ for $\mathbb{K} = \mathbb{C}$, $\mathfrak{sp}(2sn,\mathbb{R})$ for $\mathbb{K} = \mathbb{H}$.

To see that the subalgebra $\mathbb{R}[W^*]^H$ of real invariant polynomials is generated by its homogeneous degree 2 part $\mathbb{R}_2[W^*]^H$, we complexify the H-action on W and examine the three cases separately.
Case (1): $W = \mathrm{Hom}_{\mathbb{R}}(\mathbb{R}^s, \mathbb{R}^{2\ell})$, $\mathrm{Sp}(W) = \mathrm{Sp}(s\ell,\mathbb{R})$, $H = O(s', s'', \mathbb{R})$, $G = \mathrm{Sp}(\ell,\mathbb{R})$. By construction, $W_{\mathbb{C}} = W \otimes_{\mathbb{R}} \mathbb{C} = \mathbb{C}^{2s\ell} = \mathbb{C}^s \otimes_{\mathbb{C}} \mathbb{C}^{2\ell}$, that is, $W_{\mathbb{C}}$ may be written as the product of 2ℓ copies of $\mathbb{C}^s$, and the H-action on W complexifies to the diagonal action of $\mathrm{O}(s,\mathbb{C}) = H^{\mathbb{C}}$ on the product of 2ℓ copies of $\mathbb{C}^s$.
Case (2): $W = \mathrm{Hom}_{\mathbb{C}}(\mathbb{C}^s, \mathbb{C}^n) \cong \mathbb{C}^{sn}$, $\mathrm{Sp}(W) = \mathrm{Sp}(sn,\mathbb{R})$, $H = \mathrm{U}(s', s'')$, $G = \mathrm{U}(p, q)$, $n = p + q$. Now $W_{\mathbb{C}} = W \otimes_{\mathbb{R}} \mathbb{C} \cong \mathbb{C}^{2sn} \cong \mathbb{C}^{2s} \otimes_{\mathbb{C}} \mathbb{C}^n$, that is, $W_{\mathbb{C}}$ may

be written as the product of n copies of $\mathbb{C}^{2s}$, and the H-action on W complexifies to the diagonal action of $\mathrm{GL}(s,\mathbb{C}) = H^{\mathbb{C}}$ on the product of n copies of $\mathbb{C}^s \times (\mathbb{C}^s)^*$ where $(\mathbb{C}^s)^*$ is identified with $\overline{\mathbb{C}^s}$ by means of the hermitian form.
Case (3): $W = \mathrm{Hom}_{\mathbb{H}}(\mathbb{H}^s, \mathbb{H}^n) \cong \mathbb{H}^{sn}, \mathrm{Sp}(W) = \mathrm{Sp}(2sn, \mathbb{R}), H = \mathrm{Sp}(s', s''), G = \mathrm{O}^*(2n)$. Now $W_{\mathbb{C}} = W \otimes_{\mathbb{R}} \mathbb{C} \cong \mathbb{C}^{2sn} \cong \mathbb{C}^{2s} \otimes_{\mathbb{C}} \mathbb{C}^n$, that is, $W_{\mathbb{C}}$ may be written as the product of n copies of $\mathbb{C}^{2s}$, and the H-action on W complexifies to the diagonal action of $\mathrm{Sp}(s,\mathbb{C}) = H^{\mathbb{C}}$ on the product of n copies of $\mathbb{C}^{2s}$.

By the first main theorem of invariant theory, the algebra $\mathbb{C}[W^*_{\mathbb{C}}]^{H^{\mathbb{C}}}$ of complex $H^{\mathbb{C}}$-invariants in $\mathbb{C}[W^*_{\mathbb{C}}]$ is generated by its homogeneous degree 2 part $\mathbb{C}_2[W^*_{\mathbb{C}}]^{H^{\mathbb{C}}}$. Since H and G constitute a reductive dual pair in $\mathrm{Sp}(W)$, the homogeneous degree 2 part $\mathbb{C}_2[W^*_{\mathbb{C}}]^{H^{\mathbb{C}}}$ is the complexification of the homogeneous degree 2 part $\mathbb{R}_2[W^*]^H$. This implies that the subalgebra $\mathbb{R}[W^*]^H$ of real invariant polynomials is generated by its homogeneous degree 2 part $\mathbb{R}_2[W^*]^H$. Interchanging the roles of H and G we see that the subalgebra $\mathbb{R}[W^*]^G$ of real invariant polynomials is generated by its homogeneous degree 2 part $\mathbb{R}_2[W^*]^G$. □

REMARK. Our description gives the "Second Main Theorem of Invariant Theory" [**93**] for free; indeed, this theorem provides defining relations among those generators of the invariants which occur above as the components of the corresponding momentum mapping, and these defining relations just fix the rank of the matrices whose entries are the generators.

REMARK. In case (2), letting $s = 1$ so that $H = \mathrm{U}(1)$, the resulting injection of $\mathfrak{g} = \mathfrak{u}(p,q) = \mathbb{R}_2[W^*]^H$ into $\mathbb{R}_2[W^*] = \mathfrak{sp}(p+q, \mathbb{R})$ is one of the kind (3.5.1.5), up to the central factor in $\mathfrak{u}(p,q)$. Likewise, in case (3), letting $s = 1$ so that $H = \mathrm{Sp}(1)$, the resulting injection of $\mathfrak{g} = \mathfrak{so}^*(2n) = \mathbb{R}_2[W^*]^H$ into $\mathbb{R}_2[W^*] = \mathfrak{sp}(2n, \mathbb{R})$ is one of the kind (3.5.1.7).

We will now exploit Theorem 5.3.1 to relate certain reduced spaces with nilpotent orbits. The basic observation is that, under the circumstances of that theorem, given $\alpha\colon V^s \to V$, the vanishing of $\mu_H(\alpha)$ entails $\mu_G(\alpha)\mu_G(\alpha) = 0$, that is, $\mu_G(\alpha)$ is then a square zero nilpotent matrix in $\mathfrak{g}$. Since α has rank at most s, the matrix $\alpha\alpha^\dagger \in \mathfrak{g}$, viewed as a $\mathbb{K}$-linear map from $\mathbb{K}^n$ to $\mathbb{K}^n$, has $\mathbb{K}$-rank at most s. As a side remark we note that, likewise, $\mu_G(\alpha) = 0$ entails $\mu_H(\alpha)\mu_H(\alpha) = 0$; in particular, when H is compact, $\mu_H(\alpha)$ is then necessarily zero. As before, we take W^{red} to be the space of *closed* H-orbits in the zero locus of μ_H. Whether or not H is compact, by H-invariance, the momentum mapping μ_G induces a map from the space of closed H-orbits in W to $\mathfrak{g}$ and hence a map $\overline{\mu}_G\colon W^{\mathrm{red}} \to \mathfrak{g}$ the image of which, in turn, lies in the subspace of square zero nilpotent matrices in $\mathfrak{g}$ which have rank at most s since for any $\alpha\colon V^s \to V$ the image $\mu_G(\alpha) = \alpha\alpha^\dagger\colon V \to V$ has rank at most s. We will show that $\overline{\mu}_G$ identifies the H-reduced space W^{red} with the closure of finitely many pseudoholomorphic nilpotent orbits in $\mathfrak{g}$ of rank $s_r = \min(r,s)$. In particular, when H is compact, i. e. $s = s'$ and $s'' = 0$, $\overline{\mu}_G$ identifies W^{red} with the closure of the holomorphic nilpotent orbit $\mathcal{O}_{s_r}$ which consists of non-negative nilpotent matrices in $\mathfrak{g}$ of rank $s_r = \min(r,s)$.

Recall that V carries the complex structure $J = J_V$ which is induced by the element $J_V \in \mathfrak{g}$. (We remind the reader that $\frac{1}{2}J_V$ is the requisite H-element in the standard cases, cf. (3.5)). Define $J_W\colon W \to W$ by $J_W(\alpha) = J_V \circ \alpha$, where $\alpha\colon V^s \to V$. This yields a complex structure J_W on W which, together with the symplectic structure ω_W, turns W into a flat (not necessarily positive) Kähler manifold. When we wish to emphasize in notation that W is endowed with this complex structure we write W_J.

Let $V^{s-1} = \mathbb{K}^{s-1} \subseteq V^s$, and endow it with the induced hermitian form which we still write as $(\cdot,\cdot)$. Let $W(s-1) = \mathrm{Hom}_{\mathbb{K}}(V^{s-1}, V)$, $H(s-1) = \mathrm{U}(s-1,\mathbb{K})$. The inclusion $V^{s-1} \subseteq V^s$ induces a projection $V^s \to V^{s-1}$ of hermitian $\mathbb{K}$-vector spaces which, in turn, induces an injection $W(s-1) \to W = W(s)$ of $\mathbb{K}$-vector spaces compatible with all the structure: It is $H(s-1)$-linear where $H(s-1)$ is viewed as a subgroup of $H(s)$ in the obvious fashion; it is G-linear; it is compatible with the complex structure J_W; and it is compatible with the momentum mappings in the sense that the diagrams

$$\begin{array}{ccc} W(s-1) & \longrightarrow & W(s) \\ {\scriptstyle \mu_{H(s-1)}}\big\downarrow & & \big\downarrow{\scriptstyle \mu_{H(s)}} \\ \mathfrak{h}(s-1) & \longrightarrow & \mathfrak{h}(s) \end{array} \qquad \begin{array}{ccc} W(s-1) & \longrightarrow & W(s) \\ {\scriptstyle \mu_G}\big\downarrow & & \big\downarrow{\scriptstyle \mu_G} \\ \mathfrak{g} & \xrightarrow{\mathrm{Id}} & \mathfrak{g} \end{array}$$

are commutative.

We now suppose that H is compact. Thus $H = \mathrm{O}(s,\mathbb{R})$, $H = \mathrm{U}(s)$, or $H = \mathrm{Sp}(s)$, according to the case under consideration. The complex structure $J = J_W$ on W now combines with the symplectic structure ω_W to a *positive* flat Kähler structure on W—maintaining notation established earlier, we write W_J for W, endowed with this Kähler structure—, and the H-representation on W_J is unitary; in particular, it extends to a holomorphic action of $H^{\mathbb{C}}$ on W_J. Thus, in view of Proposition 4.2, via the Kempf-Ness homeomorphism from $W^{\mathrm{red}} = \mu_H^{-1}/H$ onto the categorical quotient $W_J//H^{\mathbb{C}}$, the H-reduced space W^{red} inherits a (positive) normal Kähler structure.

The next result establishes the existence of a compact associated representation for any holomorphic nilpotent orbit in a standard simple Lie algebra of hermitian type. As a by-product, independently of the classification of pseudoholomorphic nilpotent orbits given in (3.3) above, we then deduce the holomorphicity of the nilpotent orbits spelled out there. Thus the next result does *not* rely on the fact, obtained in (3.3) via the classification of pseudoholomorphic nilpotent orbits, that non-negative nilpotent elements in $\mathfrak{g}$ (i. e. matrices X in $\mathfrak{g}$ with $\mathcal{B}_X$ non-negative, see (3.5) above) generate holomorphic nilpotent orbits.

THEOREM 5.3.3. *The induced map $\overline{\mu}_G$ from the H-reduced space $W(s)^{\mathrm{red}} = \mu_H^{-1}(0)/H$ to $\mathfrak{g}$ is a proper embedding of $W(s)^{\mathrm{red}}$ into $\mathfrak{g}$ and induces an isomorphism of stratified symplectic spaces from $W(s)^{\mathrm{red}}$ onto the closure $\overline{\mathcal{O}_{s_r}}$ of the unique nilpotent orbit $\mathcal{O}_{s_r}$ which consists of non-negative nilpotent matrices in $\mathfrak{g}$ of rank $s_r = \min(r,s)$, and $\overline{\mathcal{O}_{s_r}}$ consists of non-negative nilpotent matrices in $\mathfrak{g}$ of rank at most* $\min(r,s)$. *The orbit $\mathcal{O}_{s_r}$ is necessarily holomorphic, and the isomorphism $\overline{\mu}_G$ from $W(s)^{\mathrm{red}}$ onto $\overline{\mathcal{O}_{s_r}}$ is one of complex analytic stratified Kähler spaces whence $\overline{\mathcal{O}_{s_r}}$ is a normal Kähler space. Furthermore, when $s > r$, the inclusion of $W(s-1)$ into $W(s)$ induces an isomorphism of $W(s-1)^{\mathrm{red}}$ onto $W(s)^{\mathrm{red}}$ while when $s \leq r$, so that $s_r = s$, the inclusion of $W(s-1)$ into $W(s)$ induces an injection of $W(s-1)^{\mathrm{red}}$ into $W(s)^{\mathrm{red}}$ which, under the identifications with the closures of holomorphic nilpotent orbits, amounts to the inclusion of $\overline{\mathcal{O}_{s-1}}$ into $\overline{\mathcal{O}_s}$ in such a way that $\overline{\mathcal{O}_{s-1}}$ is the complement of the top stratum $\mathcal{O}_s$ of $\overline{\mathcal{O}_s}$. Finally, under the projection from $\mathfrak{g} = \mathfrak{k} \oplus \mathfrak{p}$ to $\mathfrak{p}$, followed by the identification of the latter with $\mathfrak{p}^+$, the image of the reduced space W^{red} in $\mathfrak{g}$ is identified with the affine complex variety consisting of* *matrices in $\mathfrak{p}^+$ of rank at most s in such a way that the complex structure on $\overline{\mathcal{O}_s}$ which arises from the complex algebraic quotient construction gets identified with that resulting from the projection of $\overline{\mathcal{O}_s}$ to $\mathfrak{p}$; in particular, when $s = r$, the image of the reduced space W^{red} in $\mathfrak{g}$ is identified with $\mathfrak{p}^+$, that is, the categorical quotient $W_J//H^{\mathbb{C}}$ is identified with $\mathfrak{p}^+$.*

Thus we see once more that, for $s \leq r$, $\{0\}, \mathcal{O}_1, \ldots, \mathcal{O}_s$ are the strata of $\overline{\mathcal{O}_s}$, and that their closures constitute a chain of inclusions $\{0\} \subseteq \overline{\mathcal{O}_1} \subseteq \cdots \subseteq \overline{\mathcal{O}_{s-1}} \subset \overline{\mathcal{O}_s}$. A somewhat more explicit description of the complex analytic structures of the strata was given in Theorem 3.5.5 above. For the special case where $H = \mathrm{O}(s, \mathbb{R})$ and $G = \mathrm{Sp}(\ell, \mathbb{R})$ (case (1) above), the observation that $\overline{\mu}_G$ identifies the reduced space with the closure of a certain nilpotent orbit may be found in [**60**].

PROOF. We have already observed that the momentum mapping μ_G induces a map $\overline{\mu}_G$ from W^{red} to the space of nilpotent matrices in $\mathfrak{g}$ which have rank at most s. Since H and G centralize each other in $\mathrm{Sp}(W) = \mathrm{Sp}(\frac{dsn}{2}, \mathbb{R})$, W^{red} inherits a G-action, and $\overline{\mu}_G$ is G-equivariant. By Theorem 5.3.1, the algebra $\mathbb{R}[W^*]^H$ of H-invariants is generated by the components of $\mu_G \colon W \to \mathfrak{g}$. Hence the map from the real categorical quotient $W//H$ to $\mathfrak{g}$ induced by μ_G is the ordinary (real) Hilbert map; since for $X = \alpha\alpha^\dagger$ and $v, w \in V$,

$$\mathcal{B}_X(v, w) = -\mathcal{B}(\alpha\alpha^\dagger v, w) = \overline{\mathcal{B}(w, \alpha\alpha^\dagger v)} = \overline{(\alpha^\dagger w, \alpha^\dagger v)} = (\alpha^\dagger v, \alpha^\dagger w),$$

this Hilbert map realizes W/H (in a G-equivariant fashion) as the space of non-negative matrices in $\mathfrak{g}$ of rank at most s. Consequently the induced map $\overline{\mu}_G$ from the H-reduced space $W^{\mathrm{red}} = \mu_H^{-1}(0)/H$ to $\mathfrak{g}$ is injective as well. Furthermore, the square of the distance to the origin in W is a H-invariant polynomial function. Hence the Hilbert map is proper, and $\overline{\mu}_G$ is a G-equivariant homeomorphism from W^{red} onto its image.

Let $\mathcal{O}_{s_r}$ be the space of non-negative nilpotent matrices in $\mathfrak{g}$ of rank $s_r = \min(r, s)$. We claim that $\mathcal{O}_{s_r}$ is a single G-orbit, that $\overline{\mu}_G$ maps W^{red} onto $\overline{\mathcal{O}_{s_r}}$, and that $\overline{\mathcal{O}_{s_r}}$ is the space of non-negative nilpotent matrices in $\mathfrak{g}$ of rank at most $s_r = \min(r, s)$. To justify this claim we observe first that, for any $\alpha \colon V^s \to V$, the rank of $\alpha\alpha^\dagger$ is at most r, whatever s. Indeed, the H-reduced space W^{red} consists of orbits of $\alpha \colon V^s \to V$ such that $\alpha^\dagger\alpha \colon V^s \to V^s$ is zero; for such an α, necessarily $\alpha\alpha^\dagger\alpha = 0$, that is, $\alpha(V^s)$ lies in the kernel of $\alpha\alpha^\dagger$. Thus, if $s > r$, if $\alpha^\dagger\alpha = 0$, and if $\mathrm{rank}(\alpha) > r$, the kernel of $\alpha\alpha^\dagger$ contains a space of dimension $> r$ whence $\mathrm{rank}(\alpha\alpha^\dagger) < r$. Consequently $\overline{\mu}_G$ maps W^{red} into the space of nilpotent orbits in $\mathfrak{g}$ of rank at most $s_r = \min(r, s)$. Furthermore, this observation entails that, when $s > r$, the inclusion of $W(s-1)$ into $W(s)$ induces an isomorphism of $W(s-1)^{\mathrm{red}}$ onto $W(s)^{\mathrm{red}}$.

Next we show that, for some α_0 with $\mu_H(\alpha_0) = 0$, the image $\mu_G(\alpha_0) \in \mathfrak{g}$ has rank $s_r = \min(r, s)$. To simplify the exposition, we will henceforth suppose that $s \leq r$.

Case (1): $W = \mathrm{Hom}_{\mathbb{R}}(\mathbb{R}^s, \mathbb{R}^{2\ell})$, $\mathrm{Sp}(W) = \mathrm{Sp}(s\ell, \mathbb{R})$, $H = \mathrm{O}(s, \mathbb{R})$, $G = \mathrm{Sp}(\ell, \mathbb{R})$, $r = \ell$. Pick vectors $\mathbf{q}_1, \ldots, \mathbf{q}_\ell$ in $\mathbb{R}^s$ and consider the linear map $\alpha \colon \mathbb{R}^s \to \mathbb{R}^{2\ell}$ whose matrix has the $\mathbf{q}_1, \ldots, \mathbf{q}_\ell$ as first ℓ row vectors and whose other row vectors are zero. Then $\alpha^\dagger \colon \mathbb{R}^{2\ell} \to \mathbb{R}^s$ is given by the matrix which has the first ℓ column vectors zero and which has, for $1 \leq j \leq \ell$, the vector $\mathbf{q}_j$ as $(\ell + j)$'th column vector. Then

$\mu_H(\alpha) = \alpha^\dagger\alpha = 0$ and $\mu_G(\alpha) = \alpha\alpha^\dagger = \begin{bmatrix} 0 & Q \\ 0 & 0 \end{bmatrix}$ where $Q = [\mathbf{q}_j\mathbf{q}_k]_{1\leq j,k\leq \ell}$. Thus, when we choose $\mathbf{q}_1, \ldots, \mathbf{q}_s$ orthonormal and $\mathbf{q}_{s+1}, \ldots, \mathbf{q}_\ell$ zero and write α_0 for the resulting linear map from $\mathbb{R}^s$ to $\mathbb{R}^{2\ell}$, the image $\mu_G(\alpha_0)$ is a non-negative nilpotent matrix in $\mathfrak{sp}(\ell, \mathbb{R})$ of rank s.
Case (2): $W = \mathrm{Hom}_\mathbb{C}(\mathbb{C}^s, \mathbb{C}^n) \cong \mathbb{C}^{sn}$, $\mathrm{Sp}(W) = \mathrm{Sp}(sn, \mathbb{R})$, $H = \mathrm{U}(s)$, $G = \mathrm{U}(p, q)$, $p + q = n, p \geq q = r$. Pick vectors $\mathbf{w}_1, \ldots, \mathbf{w}_q$ in $\mathbb{C}^s$ and consider the linear map $\alpha\colon \mathbb{C}^s \to \mathbb{C}^n$ given by the transpose of the matrix

$$[\mathbf{w}_1, \ldots, \mathbf{w}_q, 0 \ldots 0, \mathbf{w}_{p+1}, \ldots, \mathbf{w}_n]$$

where, for $1 \leq j \leq q$, $\mathbf{w}_{p+j} = i\mathbf{w}_j$. Then

$$\alpha^\dagger = i[\overline{\mathbf{w}_1}, \ldots, \overline{\mathbf{w}_q}, 0, \ldots, 0, -i\overline{\mathbf{w}_1}, \ldots, -i\overline{\mathbf{w}_q}],$$

$\mu_H(\alpha) = \alpha^\dagger\alpha = 0$, and

$$\mu_G(\alpha) = \alpha\alpha^\dagger = \begin{bmatrix} iA & 0 & -A \\ 0 & 0 & 0 \\ -A & 0 & -iA \end{bmatrix}$$

where $A = [\mathbf{w}_j\overline{\mathbf{w}_k}]_{1\leq j,k\leq q}$. Thus, when we choose $\mathbf{w}_1, \ldots, \mathbf{w}_s$ orthonormal and $\mathbf{w}_{s+1}, \ldots, \mathbf{w}_q$ zero and write α_0 for the resulting linear map from $\mathbb{C}^s$ to $\mathbb{C}^{2\ell}$, the image $\mu_G(\alpha_0)$ in $\mathfrak{u}(p, q)$ is a non-negative nilpotent matrix in $\mathfrak{u}(p, q)$ of rank s.
Case (3): $W = \mathrm{Hom}_\mathbb{H}(\mathbb{H}^s, \mathbb{H}^n) \cong \mathbb{H}^{sn}, \mathrm{Sp}(W) = \mathrm{Sp}(2sn, \mathbb{R}), H = \mathrm{Sp}(s), G = \mathrm{O}^*(2n)$. The reasoning is essentially the same as in the two other cases. We leave the details to the reader.

Write $\mathcal{O}'_s$ for the G-orbit in $\mathfrak{g}$ of $\mu_G(\alpha_0)$; by construction, $\mathcal{O}'_s$ is contained in the space $\mathcal{O}_s$ of non-negative nilpotent matrices in $\mathfrak{g}$ of rank s $(\leq r)$, and the reasoning given so far implies that the closure $\overline{\mathcal{O}'_s}$ lies in the image $\overline{\mu}_G(W^{\mathrm{red}})$ of W^{red} in $\mathfrak{g}$. Thus $\overline{\mathcal{O}'_s} \subseteq \overline{\mu}_G(W^{\mathrm{red}}) \subseteq \overline{\mathcal{O}_s}$. We claim that $\mathcal{O}'_s = \mathcal{O}_s$. This follows for example from the classification of holomorphic nilpotent orbits established in (3.3) above. However, it may be verified directly, by means of an argument in the proof of Theorem 5.3.1 of [**60**]: Suppose for the moment that $\mathfrak{g} = \mathfrak{sp}(\ell, \mathbb{R})$ $(r = \ell)$ and let X be a non-negative nilpotent matrix in $\mathfrak{g}$ of rank s $(\leq r)$. Then X is of square zero and hence, when X is viewed as an endomorphism of V $(= \mathbb{R}^{2\ell})$, the image $\mathrm{im}(X)$ of X is $\mathcal{B}$-isotropic, since $\mathcal{B}(Xu, Xv) = \mathcal{B}(u, X^2v)$. Furthermore, $\mathrm{im}(X) \subseteq \ker(X)$. Write $V = L_1 \oplus L_2$ as a direct sum of transverse Lagrangian subspaces L_1 and L_2, with $\mathrm{im}(X) \subseteq L_1$, and pick $A \in \mathrm{GL}(L_1)$ such that $B = \begin{bmatrix} A & 0 \\ 0 & (A^t)^{-1} \end{bmatrix} \in \mathrm{Sp}(V)$ carries the decomposition $V = L_1 \oplus L_2$ into the standard decomposition $\mathbb{R}^{2\ell} = \mathbb{R}^\ell \oplus \mathbb{R}^\ell$. Then BXB^{-1} has the form $Y = \begin{bmatrix} 0 & S \\ 0 & 0 \end{bmatrix}$ where S is a symmetric $(\ell \times \ell)$-matrix of rank s. Thus we may assume that X is of the kind Y. Furthermore, when we choose B suitably, the matrix S will be a diagonal matrix. Hence the orbit of X is determined by the signature of S. The positivity assumption means that the signature of S is zero. The matrix Y plainly lies in $\mathcal{O}'_s$ whence $\overline{\mathcal{O}_s} \subseteq \overline{\mathcal{O}'_s}$. This kind of reasoning may be applied to the other cases as well. Alternatively, we may reduce these cases to that of $\mathfrak{sp}(\ell, \mathbb{R})$ (for suitable ℓ) by means of the embeddings (3.5.1.5) and (3.5.1.7). A closer look shows that the argument comes down to Gram-Schmidt orthogonalization of the complex subspace of V generated by the image of X, with respect to the complex structure J_V.

Finally we examine the complex analytic structures.
Case (1): The Cartan decomposition $\mathfrak{sp}(\ell,\mathbb{R}) = \mathfrak{u}(\ell)\oplus\mathfrak{p}$ of $\mathfrak{sp}(\ell,\mathbb{R})$ complexifies to $\mathfrak{sp}(\ell,\mathbb{C}) = \mathfrak{gl}(\ell,\mathbb{C})\oplus\mathfrak{p}^+\oplus\mathfrak{p}^-$, where $\mathfrak{p}^+ = \mathrm{S}^2_{\mathbb{C}}[\mathbb{C}^\ell]$ and $\mathfrak{p}^- = \overline{\mathrm{S}^2_{\mathbb{C}}[\mathbb{C}^\ell]}$. Being endowed with the complex structure $J = J_V$ on $W \cong \mathbb{R}^{2s\ell}$, this space is just a copy of $\mathbb{C}^{s\ell}$, $H^{\mathbb{C}} = \mathrm{O}(s,\mathbb{C})$, and the $H^{\mathbb{C}}$-action on $W_J = \mathbb{C}^{s\ell}$ amounts to the standard diagonal action of $\mathrm{O}(s,\mathbb{C})$ on a product $\mathbb{C}^s\times\cdots\times\mathbb{C}^s$ of ℓ copies of $\mathbb{C}^s$. By the first main theorem of invariant theory for the group $\mathrm{O}(s,\mathbb{C})$, for $s\geq 1$, the invariants in $\mathbb{C}[W_J^*]^{H^{\mathbb{C}}}$ are generated by the homogeneous degree 2 part $\mathbb{C}_2[W_J^*]^{H^{\mathbb{C}}}$ thereof, and the complex algebraic map—the (complex) Hilbert map—from $W_J = (\mathbb{C}^s)^\ell$ to the complex vector space $\mathrm{S}^2_{\mathbb{C}}[\mathbb{C}^\ell]$ of complex symmetric $(\ell\times\ell)$-matrices which assigns the symmetric matrix $[\mathbf{w}_j\mathbf{w}_k]_{1\leq j,k\leq\ell}$ to $\mathbf{w} = (\mathbf{w}_1,\dots,\mathbf{w}_\ell)\in W_J$ identifies the complex algebraic quotient $W_J//\mathrm{O}(s,\mathbb{C})$ and hence the reduced space W^{red} as a complex analytic space with the affine complex variety consisting of matrices in $\mathrm{S}^2_{\mathbb{C}}[\mathbb{C}^\ell]$ of rank at most s.
Case (2): The Cartan decomposition $\mathfrak{u}(p,q) = \mathfrak{u}(p)\oplus\mathfrak{u}(q)\oplus\mathfrak{p}$ of $\mathfrak{u}(p,q)$ has

$$\mathfrak{p} = \mathrm{M}_{q,p}(\mathbb{C})$$

and complexifies to

$$(\mathfrak{u}(p,q))^{\mathbb{C}} = \mathfrak{gl}(n,\mathbb{C}) = \mathfrak{gl}(p,\mathbb{C})\oplus\mathfrak{gl}(q,\mathbb{C})\oplus\mathfrak{p}^+\oplus\mathfrak{p}^-,$$

where $\mathfrak{p}^+ = \mathrm{M}_{q,p}(\mathbb{C})$ and $\mathfrak{p}^- = \overline{\mathrm{M}_{q,p}(\mathbb{C})}\cong\mathrm{M}_{p,q}(\mathbb{C})$. Being endowed with the complex structure $J = J_V$ on $W\cong\mathbb{C}^{sn}$, this space is a copy of $\mathbb{C}^{sn}$, $H^{\mathbb{C}} = \mathrm{GL}(s,\mathbb{C})$, and the $H^{\mathbb{C}}$-action on $W_J = \mathbb{C}^{sn}$ amounts to the standard diagonal action of $\mathrm{GL}(s,\mathbb{C})$ on a product

$$\mathbb{C}^s\times\cdots\times\mathbb{C}^s\times(\mathbb{C}^s)^*\times\cdots\times(\mathbb{C}^s)^*$$

of p copies of $\mathbb{C}^s$ and q copies of $(\mathbb{C}^s)^*$. By invariant theory, just as before, the invariants in $\mathbb{C}[W_J^*]^{H^{\mathbb{C}}}$ are generated by the homogeneous degree 2 part $\mathbb{C}_2[W_J^*]^{H^{\mathbb{C}}}$ thereof, and the Hilbert map from W_J to $\mathrm{M}_{q,p}(\mathbb{C})$ assigns the matrix $[(\xi_j,\mathbf{w}_k)]_{1\leq j\leq q,1\leq k\leq p}$ to $\mathbf{w} = (\mathbf{w}_1,\dots,\mathbf{w}_p,\xi_1,\dots,\xi_q)\in W_J$ and thereby identifies the complex algebraic quotient $W_J//\mathrm{GL}(s,\mathbb{C})$ and hence the reduced space W^{red} as a complex analytic space with the complex affine variety in $\mathrm{M}_{q,p}(\mathbb{C})$ which consists of matrices of rank at most $\min(s,q)$. In particular, when $s\geq q$, there are no relations among the invariants; indeed, any complex $(p\times q)$-matrix may be written as $[(\xi_j,\mathbf{w}_k)]_{1\leq j\leq q,1\leq k\leq p}$ for suitable vectors $\mathbf{w}_1,\dots,\mathbf{w}_p\in\mathbb{C}^s$ and $\xi_1,\dots,\xi_q\in(\mathbb{C}^s)^*$ provided $s\geq q$. Consequently, when $s\geq q$, the algebraic quotient $W_J//H^{\mathbb{C}}$ actually coincides with $\mathfrak{p}^+ = \mathrm{M}_{q,p}(\mathbb{C})$.
Case (3): The Cartan decomposition $\mathfrak{so}^*(2n) = \mathfrak{u}(n)\oplus\mathfrak{p}$ of $\mathfrak{so}^*(2n)$ has

$$\mathfrak{p} = i\Lambda^2_{\mathbb{R}}[\mathbb{R}^n]\oplus i\Lambda^2_{\mathbb{R}}[\mathbb{R}^n]$$

and complexifies to

$$(\mathfrak{so}^*(2n))^{\mathbb{C}} = \mathfrak{so}(2n,\mathbb{C}) = \mathfrak{gl}(n,\mathbb{C})\oplus\mathfrak{p}^+\oplus\mathfrak{p}^-$$

where $\mathfrak{p}^+ = \Lambda^2_{\mathbb{C}}[\mathbb{C}^n]$ and $\mathfrak{p}^- = \overline{\Lambda^2_{\mathbb{C}}[\mathbb{C}^n]}$. With respect to the complex structure $J = J_V$ on $W\cong\mathbb{H}^{sn}$, the latter is a copy of $\mathbb{C}^{2sn}$, $H^{\mathbb{C}} = \mathrm{Sp}(s,\mathbb{C})$, and the $H^{\mathbb{C}}$-action on $W_J = \mathbb{C}^{2sn}$ amounts to the standard diagonal action of $\mathrm{Sp}(s,\mathbb{C})$ on a product $\mathbb{C}^{2s}\times\cdots\times\mathbb{C}^{2s}$ of n-copies of $\mathbb{C}^{2s}$. By invariant theory, just as before, the invariants in $\mathbb{C}[W_J^*]^{H^{\mathbb{C}}}$ are generated by the homogeneous degree 2 part $\mathbb{C}_2[W_J^*]^{H^{\mathbb{C}}}$ thereof, and from the theory of reductive dual pairs we know that $\mathbb{C}_2[W_J^*]^{H^{\mathbb{C}}}$ is a copy of $\mathfrak{so}(n,\mathbb{C}) \cong \Lambda^2_{\mathbb{C}}[\mathbb{C}^n] = \mathfrak{p}^+$. With reference to the notation ω for the standard complex symplectic form on $\mathbb{C}^{2s}$, the Hilbert map from W_J to $\Lambda^2_{\mathbb{C}}[\mathbb{C}^n]$ assigns the skew-symmetric complex $(n \times n)$-matrix $[\omega(\mathbf{w}_j, \mathbf{w}_k)]_{1 \leq j,k \leq n}$ to $(\mathbf{w}_1, \dots, \mathbf{w}_n) \in W_J$ and thereby identifies the complex algebraic quotient $W_J//H^{\mathbb{C}}$, and hence the reduced space W^{red} as a complex analytic space, with a complex affine variety in $\Lambda^2_{\mathbb{C}}[\mathbb{C}^n] = \mathfrak{so}(n,\mathbb{C})$. When $s \geq r (= [\frac{n}{2}])$, there are no relations among the invariants; indeed, any skew-symmetric complex $(n \times n)$-matrix may be written as $[\omega(\mathbf{w}_j, \mathbf{w}_k)]_{1 \leq j,k \leq n}$ for suitable vectors $\mathbf{w}_1, \dots, \mathbf{w}_n \in \mathbb{C}^{2s}$ provided $2s \geq n$. Consequently, when $s \geq r$, the algebraic quotient $W_J//H^{\mathbb{C}}$ coincides with $\mathfrak{p}^+ = \Lambda^2_{\mathbb{C}}[\mathbb{C}^n]$. When $2s < n$, the Hilbert map from W_J to $\Lambda^2_{\mathbb{C}}[\mathbb{C}^n]$ identifies the complex algebraic quotient $W_J//H^{\mathbb{C}}$ with the complex affine variety of matrices in $\Lambda^2_{\mathbb{C}}[\mathbb{C}^n] = \mathfrak{so}(n,\mathbb{C})$ having rank at most $2s$.

Thus, in each of the three standard cases, when $1 \leq s \leq r$, under the projection from $\mathfrak{g} = \mathfrak{k} \oplus \mathfrak{p}$ to $\mathfrak{p}$, followed by the identification of the latter with $\mathfrak{p}^+$, the (image of the) reduced space W^{red} in $\mathfrak{g}$ is identified with the affine complex variety consisting of matrices in $\mathfrak{p}^+$ of rank at most s for $\mathfrak{g} = \mathfrak{sp}(\ell, \mathbb{R})$ and $\mathfrak{g} = \mathfrak{su}(p,q)$ and of rank at most $2s$ for $\mathfrak{g} = \mathfrak{so}^*(2n)$; in particular, when $s = r$, the (image of the) reduced space W^{red} in $\mathfrak{g}$, that is, (that of) the categorical quotient $W_J//H^{\mathbb{C}}$ is identified with $\mathfrak{p}^+$. Furthermore, the complex structure on $\overline{\mathcal{O}_s}$ which arises from the algebraic quotient construction gets identified with that resulting from the projection of $\overline{\mathcal{O}_s}$ to $\mathfrak{p}$. Consequently, for $1 \leq s \leq r$, the restriction to the closure $\overline{\mathcal{O}'_s}$ of the nilpotent (G-)orbit $\mathcal{O}'_s$ of the projection from $\mathfrak{g} = \mathfrak{k} \oplus \mathfrak{p}$ to $\mathfrak{p}$ is injective, whence $\mathcal{O}'_s$ is a holomorphic nilpotent orbit. □

We mention in passing that the reasoning aimed at unravelling the complex analytic structures of the categorical quotients written as $W_J//H^{\mathbb{C}}$, given in the proof of Theorem 5.3.3, establishes the claim of Theorem 3.5.5 above as well.

5.4. THE REAL SEMI-ALGEBRAIC STRUCTURE. The proof of Theorem 5.3.3 actually yields more explicit information, which we now spell out. To this end, we write ρ for the (ordinary) rank of $\mathfrak{g}$. Recall that $\rho(\mathfrak{sp}(\ell,\mathbb{R})) = \ell$, $\rho(\mathfrak{u}(p,q)) = p + q = n$, $\rho(\mathfrak{so}^*(2n)) = n$. Let $\beta_1, \dots, \beta_\rho$ be a set of generators for the G-invariants on $\mathfrak{g}$ and view them as polynomial functions on $\mathfrak{g}$ as usual by means of the half-trace pairing (the appropriate multiple of the Killing form). Let $\beta = (\beta_1, \dots, \beta_\rho) \colon \mathfrak{g} \to \mathbb{R}^\rho$.

THEOREM 5.4.1. *In the standard cases, for $1 \leq s \leq r$, the real algebraic quotient $W//H$ is the space of matrices in $\mathfrak{g}$ which have rank at most s (real rank s for $\mathfrak{g} = \mathfrak{sp}(\ell,\mathbb{R})$, $\ell = r$; complex rank s for $\mathfrak{g} = \mathfrak{su}(p,q)$, $p \geq q = r$; quaternionic rank s for $\mathfrak{g} = \mathfrak{so}^*(2n)$, $r = [\frac{n}{2}]$), that is, $W//H$ is a real determinantal variety in $\mathfrak{g}$. Furthermore, the image $p(W)$ of W under the projection from W to $W//H$ sits inside $W//H$ as the space of non-negative matrices in $W//H$, and the reduced space W^{red} sits inside $p(W)$ as the zero locus of β, that is, of the G-invariants on $\mathfrak{g}$, viewed as functions on $W//H$ (and hence on $p(W)$). Finally, for $s \geq 1$, the Zariski tangent space $\mathrm{T}_0(\overline{\mathcal{O}_s})$ at the origin (for the smooth structure $C^\infty(\overline{\mathcal{O}_s})$) coincides with the entire ambient space $\mathfrak{g}$ whence $\overline{\mathcal{O}_s}$ cannot be realized in any proper linear subspace of $\mathfrak{g}$. In particular, the stratified symplectic Poisson structure on* *each* $\overline{\mathcal{O}_s}$ *necessarily involves* all *the elements of a basis of* $\mathfrak{g}$, *i. e. cannot be described by fewer generators than the dimension of* $\mathfrak{g}$.

PROOF. Since the momentum mapping $\mu_G: W \to \mathfrak{g}$ amounts to the Hilbert map for the H-action on W, μ_G induces an injection of the real algebraic quotient $W//H$ into $\mathfrak{g}$ which identifies $W//H$ with the subspace of $\mathfrak{g}$ of matrices of rank at most s (real rank s for $\mathfrak{g} = \mathfrak{sp}(\ell, \mathbb{R})$; complex rank s for $\mathfrak{g} = \mathfrak{su}(p,q)$; quaternionic rank s for $\mathfrak{g} = \mathfrak{so}^*(2n)$).

The image $p(W)$ of W in $W//H$ under the projection $p: W \to W//H \subseteq \mathfrak{g}$ is the space W/H of H-orbits in $W//H$; it is the real semi-algebraic set of non-negative matrices in $\mathfrak{g}$ of rank at most s ($\leq \ell$). Recall that a matrix X in $\mathfrak{g}$ is non-negative if the hermitian form $-\mathcal{B}(X\cdot,\cdot) = -(J_V X\cdot,\cdot)$ is non-negative or, equivalently, if the hermitian matrix $-J_V X$ is non-negative. The property of $-J_V X$ being non-negative may be expressed in terms of suitable inequalities involving the appropriate minors. Thus, as a real semi-algebraic subset of $\mathfrak{g}$, the image $p(W)$ is given by the equations which say that all $(s+1) \times (s+1)$-minors vanish and the inequalities saying that a matrix X in $\mathfrak{g}$ of rank at most s is non-negative.

Since the nilcone in $\mathfrak{g}$ is the zero locus of the G-invariants $\beta_1, \dots, \beta_\rho$ in $\mathfrak{g}$, viewed as functions on $\mathfrak{g}$, the zero locus of these invariants, viewed as functions on $W//H$, is the space of nilpotent matrices in $\mathfrak{g}$ of rank $\leq s$. Hence the zero locus of the invariants, viewed as functions on $p(W) \subseteq W//H$, is the space of non-negative nilpotent matrices in $\mathfrak{g}$ of rank $\leq s$. Thus, in view of Theorem 5.3.3, the reduced space W^{red} or, equivalently, the closure of the holomorphic nilpotent orbit $\mathcal{O}_s$ which consists of non-negative nilpotent matrices in $\mathfrak{g}$ of rank $\leq s$, sits inside $p(W)$ as the zero locus of β, that is, of the G-invariants on $\mathfrak{g}$, viewed as functions on $W//H$ (and hence on $p(W)$).

To justify the claim involving the Zariski tangent spaces at the origin, it will suffice to prove that the Zariski tangent space $\mathrm{T}_0(\overline{\mathcal{O}_1})$ at the origin (for the smooth structure $C^\infty(\overline{\mathcal{O}_1})$) coincides with the entire ambient space $\mathfrak{g}$. In case (1), for $s = 1$, $H = \mathrm{O}(1, \mathbb{R})$, the H-momentum mapping is trivial, and the G-momentum mapping $\mu_G: (\mathbb{R}^2)^\ell \to \mathfrak{sp}(\ell, \mathbb{R})$ sends the vector $[q_1, p_1, \dots, q_\ell, p_\ell]^\tau$ (where $q_j, p_j \in \mathbb{R}$) to the matrix

$$\begin{bmatrix} [q_j p_k] & -[q_j q_k] \\ [p_j p_k] & -[p_j q_k] \end{bmatrix} \in \mathfrak{sp}(\ell, \mathbb{R});$$

it is thus straightforward to find suitable vectors $[q_1, p_1, \dots, q_\ell, p_\ell]^\tau$ whose images in $\mathfrak{sp}(\ell, \mathbb{R})$ constitute a basis thereof. Likewise, in case (2), for $s = 1$, $H = \mathrm{U}(1)$, the H-momentum mapping $\mu_H: \mathbb{C}^{p+q} \to \mathfrak{u}(1)$ sends the vector $[w_1, \dots, w_{p+q}]^\tau$ (where $w_1, \dots, w_{p+q} \in \mathbb{C}$) to

$$i(w_1\overline{w_1} + \dots + w_p\overline{w_p} - w_{p+1}\overline{w_{p+1}} - \dots - w_{p+q}\overline{w_{p+q}}) \in \mathfrak{u}(1),$$

and the G-momentum mapping $\mu_G: \mathbb{C}^{p+q} \to \mathfrak{u}(p,q)$ sends $[w_1, \dots, w_{p+q}]^\tau$ to

$$\begin{bmatrix} A & B \\ B^* & D \end{bmatrix} \in \mathfrak{u}(p,q)$$

where

$$A = i\,[w_j\overline{w}_k]_{1\leq j,k\leq p}\,, \quad B = -i\,[w_j\overline{w}_{p+k}]_{1\leq j\leq p, 1\leq k\leq q}\,, \quad D = -i\,[w_{j+p}\overline{w}_{k+p}]_{1\leq j,k\leq q}\,.$$

In particular, when $\mu_H([w_1, \dots, w_{p+q}]^\tau) = 0$, the value $\mu_G([w_1, \dots, w_{p+q}]^\tau)$ lies in $\mathfrak{su}(p,q)$. A little thought reveals that, for suitable choices of $w_1, \dots, w_{p+q}$ with

$\mu_H([w_1,\dots,w_{p+q}]^\tau)=0$, the images $\mu_G([w_1,\dots,w_{p+q}]^\tau)\in\mathfrak{su}(p,q)$ constitute a basis. For example, for $\mathfrak{su}(2,1)$, the vectors $[1,0,1]^\tau$, $[0,1,1]^\tau$, $[1,0,i]^\tau$, $[0,1,i]^\tau$, $[i,0,1]^\tau$, $[0,i,1]^\tau$, $[1,1,\sqrt{2}]^\tau$, $[i,1,\sqrt{2}]^\tau$ will have the asserted properties. The same kind of reasoning establishes the claim in case (3). We leave the details to the reader. □

REMARK 5.4.2. Theorem 5.4.1 yields, for $1\leq s\leq r$, explicit equations and inequalities for the closure $\overline{\mathcal{O}_s}$, realized as a semi-algebraic set in $\mathfrak{g}$. Indeed, $\overline{\mathcal{O}_s}$ consists of all matrices $\alpha\in\mathfrak{g}$ (taken in its matrix realization spelled out in (3.5)) which satisfy

1. the real determinantal equations which say that all $((s+1)\times(s+1))$-minors of α vanish;
2. the equations $\beta_1(\alpha)=0,\dots,\beta_\rho(\alpha)=0$ (which say that α lies in the real nilvariety of $\mathfrak{g}$), and
3. the inequalities which say that the matrix α is non-negative.

REMARK 5.4.3. Theorems 5.3.3 and 5.4.1 show once more that, for a standard simple Lie algebra of hermitian type $(\mathfrak{g},z)$ of real rank r, even though as a complex analytic space, the closure $\overline{\mathcal{O}_r}$ of the top stratum is affine and hence non-singular, as a stratified symplectic space, it has *singularities*, the *singular locus* (in the sense of stratified symplectic spaces) being the complement $\overline{\mathcal{O}_{r-1}}$ of the top stratum $\mathcal{O}_r$. Furthermore, Theorem 5.3.3 entails at once the fact, implied as well by Corollary 3.3.8 above that, for a holomorphic nilpotent orbit $\mathcal{O}$, the diffeomorphism from $\mathcal{O}$ onto its image in $\mathfrak{p}$ extends to a homeomorphism from the closure $\overline{\mathcal{O}}$ onto its image in $\mathfrak{p}$.

COROLLARY 5.4.4. *Let H be a compact Lie group which decomposes into a direct product of copies of* $\mathrm{O}(s_j,\mathbb{R})$, $\mathrm{Sp}(s_k)$, $\mathrm{U}(s_m)$, *and let E be a representation of H which decomposes into a product of standard representations and of the conjugate representation of the standard one for the case of* $\mathrm{U}(s_m)$. *Then the reduced space E^{red} is as a normal Kähler space isomorphic to a holomorphic nilpotent orbit in a real reductive Lie algebra $(\mathfrak{g},z)$ of hermitian type which decomposes into a sum of copies of* $\mathfrak{sp}(\ell_j,\mathbb{R})$, $\mathfrak{so}^*(2\ell_k)$, $\mathfrak{u}(p_m,q_m)$, *for suitable* ℓ_j,ℓ_k,p_m,q_m. *Furthermore, given a real reductive Lie algebra of hermitian type which decomposes into a sum of Lie algebras of the standard kind, every holomorphic nilpotent orbit in such a Lie algebra arises in this fashion.*

PROOF. This is an immediate consequence of Theorem 5.4.1. □

5.5. EXISTENCE OF ASSOCIATED REPRESENTATIONS FOR *arbitrary* PSEUDOHOLOMORPHIC NILPOTENT ORBITS. Return to the real symplectic vector space $W=W(s)=\mathrm{Hom}_{\mathbb{K}}(V^s,V)$, endowed with the symplectic actions of the groups H and G and corresponding momentum mappings where H is no longer assumed to be compact. As before, write (s',s'') for the type of the hermitian form $(\cdot,\cdot)$ on W, so that $s'+s''$ is its rank and $s'-s''$ its signature.

THEOREM 5.5.1. *The induced map $\overline{\mu}_G$ from the H-reduced space $W(s)^{\mathrm{red}}$ to $\mathfrak{g}$ is an embedding of $W(s)^{\mathrm{red}}$ into $\mathfrak{g}$ and induces an isomorphism of stratified spaces from $W(s)^{\mathrm{red}}$ onto the union N of all pseudoholomorphic nilpotent orbits $\mathcal{O}_{(t,u)}$ in $\mathfrak{g}$ with $t\leq\min(r,s')$, $u\leq\min(r,s'')$ (and, of course, $t+u\leq r$). Furthermore, the map from $W(s)^{\mathrm{red}}$ to N is a Poisson map.*

PROOF. This is proved in much the same way as Theorem 5.3.3. We leave the details to the reader. □

Thus when $s = 2r$ and the hermitian form $(\cdot, \cdot)$ on V^s has type $(s', s'') = (r, r)$, the reduced space $W(s)^{\text{red}}$ recovers *all* pseudoholomorphic nilpotent orbits in $\mathfrak{g}$.

REMARK 5.5.2. By symmetry, the above construction identifies the G-reduced space with a union of square zero nilpotent orbits in $\mathfrak{h}$, and every square zero nilpotent orbit orbit in $\mathfrak{h}$ arises in this way. This yields the square zero nilpotent orbits for the classical Lie algebras $\mathfrak{so}(s', s'')$, $\mathfrak{u}(s', s'')$, and $\mathfrak{sp}(s', s'')$. The case $\mathfrak{u}(s', s'')$ does not provide anything new, though. The case of $\mathfrak{so}(2, q)$ will be examined in Section 6 below.

REMARK 5.5.3. Earlier we made the observation that, for a pseudoholomorphic nilpotent orbit $\mathcal{O}$ which is neither holomorphic nor antiholomorphic, the diffeomorphism from $\mathcal{O}$ onto its image in $\mathfrak{p}$ under the projection from $\mathfrak{g}$ to $\mathfrak{p}$ does not extend to a homeomorphism. This observation translates to the fact that the obvious map from the reduced space $W(s)^{\text{red}}$ to the corresponding categorical quotient is *not* a homeomorphism. *Thus the Kempf-Ness observation (relating the symplectic quotient for a hamiltonain action of a compact group with the categorical quotient for the complexified group) does not in general extend to the case where a complex group acts holomorphically with hamiltonian action of the restriction to a non-compact real form.*

COROLLARY 5.5.4. *Let H be a Lie group which decomposes into a direct product of copies of $\mathrm{O}(s'_j, s''_j)$, $\mathrm{Sp}(s'_k, s''_k)$, $\mathrm{U}(s'_m, s''_m)$, and let E be a representation of H which decomposes into a product of standard representations and of the conjugate representation of the standard one for the case of $\mathrm{U}(s'_m, s''_m)$. Then the reduced space E^{red} is as a stratified space isomorphic to a union N of pseudoholomorphic nilpotent orbits in a real reductive Lie algebra $(\mathfrak{g}, z)$ of hermitian type which decomposes into a sum of copies of $\mathfrak{sp}(\ell_j, \mathbb{R})$, $\mathfrak{so}^*(2\ell_k)$, $\mathfrak{u}(p_m, q_m)$, for suitable ℓ_j, ℓ_k, p_m, q_m. Furthermore, the map from E^{red} to N is a Poisson map, and every pseudoholomorphic nilpotent orbit in a real reductive Lie algebra of hermitian type which decomposes into a sum of Lie algebras of the standard kind arises in this fashion as a constituent of a reduced space.*

PROOF. This is again an immediate consequence of Theorem 5.4.1. □

EXAMPLE 5.5.5. Let V^1 be the standard representation of $H = \mathrm{O}(1,1)$ on $\mathbb{R}^2$ and V the standard representation of $G = \mathrm{Sp}(1, \mathbb{R})$ on $\mathbb{R}^2$. Then $W = \mathrm{Hom}_{\mathbb{R}}(V^1, V)$ has dimension four, as an H-representation, W amounts to the sum $V^1 \oplus V^1$ of two copies of the standard representation, the H-momentum mapping sends $\alpha = (\mathbf{q}, \mathbf{p}) \in \mathbb{R}^2 \times \mathbb{R}^2$ to $\mathbf{q} \wedge \mathbf{p} \begin{bmatrix} 0 & 1 \\ 1 & 0 \end{bmatrix}$, and the G-momentum mapping $\mu_G \colon W \to \mathfrak{sp}(1, \mathbb{R})$ sends $\alpha = (\mathbf{q}, \mathbf{p}) \in \mathbb{R}^2 \times \mathbb{R}^2$ to

$$\alpha\alpha^\dagger = \begin{bmatrix} (\mathbf{q}, \mathbf{p}) & -(\mathbf{q}, \mathbf{q}) \\ (\mathbf{p}, \mathbf{p}) & -(\mathbf{q}, \mathbf{p}) \end{bmatrix} \in \mathfrak{sp}(1, \mathbb{R}).$$

The zero locus $\mu_H^{-1}(0)$ consists of the points $(\mathbf{q}, \lambda\mathbf{q})$ and $(\lambda\mathbf{q}, \mathbf{q})$ ($\mathbf{q} = (q_1, q_2) \in \mathbb{R}^2$, $\lambda \in \mathbb{R}$), and the non-closed H-orbits in $\mu_H^{-1}(0)$ are those which correspond to non-zero isotropic elements $\mathbf{q}$, that is, to non-zero $\mathbf{q}$'s with $(\mathbf{q}, \mathbf{q}) = q_1^2 - q_2^2 = 0$.

The H-reduced space is the *whole* nilcone in $\mathfrak{sp}(1,\mathbb{R})$, that is, the closure of the holomorphic *and* antiholomorphic 2-dimensional nilpotent orbits in $\mathfrak{sp}(1,\mathbb{R})$.

5.6. THE SEMISIMPLE HOLOMORPHIC ORBITS. Maintaining the notation established earlier, let $s = 2\ell, p+q, n$, and $s' = s, s'' = 0$, so that, according to the case considered:

(1) $W = \mathrm{End}_{\mathbb{R}}(\mathbb{R}^{2\ell})$, $(G,H,K) = (\mathrm{Sp}(\ell,\mathbb{R}), \mathrm{O}(2\ell,\mathbb{R}), \mathrm{U}(\ell))$
(2) $W = \mathrm{End}_{\mathbb{C}}(\mathbb{C}^{p,q})$, $(G,H,K) = (\mathrm{U}(p,q), \mathrm{U}(p+q), \mathrm{U}(p)\times\mathrm{U}(q))$
(3) $W = \mathrm{End}_{\mathbb{H}}(\mathbb{H}^{n})$, $(G,H,K) = (\mathrm{O}^*(2n), \mathrm{Sp}(n), \mathrm{U}(n))$.

In particular, the group H is compact. Then J_V lies in $\mathfrak{h}$, the pre-image $\mu_H^{-1}(J_V)$ may be identified with the group G—in fact, this is the very definition of G as the group of $\mathcal{B}$-isometries of V, cf. (3.5.1) above—, the stabilizer of J_V in H may be identified with the maximal compact subgroup K of G determined by the choice of H-element, and the reduced space $\mu_H^{-1}(J_V)/K$ amounts to the homogeneous space G/K. Thus the G-momentum mapping $\mu_G\colon W \to \mathfrak{g}$ ($\mathfrak{g}$ being identified with its dual as explained earlier) identifies the reduced space $\mu_H^{-1}(J_V)/K \cong G/K$ with the semisimple holomorphic orbit $G(2z)$ of $2z$ in $\mathfrak{g}$. More generally, for $\varepsilon > 0$, the G-momentum mapping $\mu_G\colon W \to \mathfrak{g}$ identifies the reduced space $\mu_H^{-1}(\varepsilon J_V)/K$ with the semisimple holomorphic orbit $G(2\varepsilon z)$ of $2\varepsilon z$ in $\mathfrak{g}$, and this orbit is still canonically diffeomorphic to G/K.

6. Associated representations for the remaining classical case

While the construction given in Section 5 does not yield the holomorphic nilpotent orbits in $\mathfrak{so}(2,q)$ by symplectic reduction with respect to a *compact group*, these orbits may be obtained by symplectic reduction with respect to the non-compact group $\mathrm{Sp}(1,\mathbb{R})$, as already hinted at in Remark 5.9 above.

6.1. THE REAL STRUCTURE. Let $V = \mathbb{R}^n$ $(n \geq 2)$, endowed with the standard Lorentz form $(\cdot,\cdot)$ of signature (p,q) $(p+q \geq 2)$—we indicate this with the notation $\mathbb{R}^{p,q}$, so that $\mathrm{U}(V,(\cdot,\cdot)) = \mathrm{O}(p,q)$—and, for $s \geq 1$, consider $\mathbb{R}^{2s}$, endowed with its standard symplectic structure which, for consistency of exposition, we write $\mathcal{B}(\cdot,\cdot)$, so that $\mathrm{U}(\mathbb{R}^{2s},\mathcal{B}) = \mathrm{Sp}(s,\mathbb{R})$; the notation $\mathrm{U}(\cdot,\cdot)$ was introduced in (3.5) above. Let $W = W(s) = \mathrm{Hom}_{\mathbb{R}}(\mathbb{R}^{2s}, V)$. As before, the groups $H = \mathrm{Sp}(s,\mathbb{R})$ and $G = \mathrm{O}(p,q)$ act on W by means of the assignment to (x,y,α) of $y\alpha x^{-1}$ where $x \in \mathrm{Sp}(s,\mathbb{R})$, $\alpha \in \mathrm{Hom}_{\mathbb{K}}(V^s, V)$, $y \in \mathrm{O}(p,q)$. Given an $\mathbb{R}$-linear map $\alpha\colon \mathbb{R}^{2s} \to \mathbb{R}^{p,q}$, define $\alpha^\dagger\colon \mathbb{R}^{p,q} \to \mathbb{R}^{2s}$ by

$$(\alpha\mathbf{u},\mathbf{v}) = \mathcal{B}(\mathbf{u},\alpha^\dagger\mathbf{v}),\quad \mathbf{u} \in \mathbb{R}^{2s},\quad \mathbf{v} \in \mathbb{R}^{p,q};$$

we note that here $\mathcal{B}$ and $(\cdot,\cdot)$ are defined on the domain and range, respectively, of α whereas in Section 5 above $\mathcal{B}$ and $(\cdot,\cdot)$ are defined on the range and domain, respectively, of the α coming into play there. The forms $\mathcal{B}$ and $(\cdot,\cdot)$ induce a symplectic structure on $W \cong \mathbb{R}^{2sn}$, and the actions of H and G on W are hamiltonian, having momentum mappings

$$\mu_H\colon W \longrightarrow \mathfrak{h},\quad \mu_H(\alpha) = -\alpha^\dagger\alpha\colon \mathbb{R}^{2s} \to \mathbb{R}^{2s},$$
$$\mu_G\colon W \longrightarrow \mathfrak{g},\quad \mu_G(\alpha) = \alpha\alpha^\dagger\colon \mathbb{R}^{p,q} \to \mathbb{R}^{p,q}$$

respectively where, as before, $\mathfrak{h}$ and $\mathfrak{g}$ are identified with their duals by means of the half-trace pairings. The groups $\mathrm{Sp}(s,\mathbb{R})$ and $\mathrm{O}(p,q)$ constitute a dual pair in $\mathrm{Sp}(W,\mathbb{R}) \cong \mathrm{Sp}(sn,\mathbb{R})$ [**28**] (5.2(i)) and, up to sign, the H-momentum mapping μ_H amounts to the Hilbert map for the H-action on W, that is to say, it passes to an embedding of the real algebraic quotient $W//H$ into $\mathfrak{so}(p,q)$. As before, let W^{red} be the space of *closed* H-orbits in $\mu_H^{-1}(0)$, viewed as a subspace of $W//H$.

We will now unravel the real orbit structure which is relevant for our purposes: Let $s = 1$, $p = 2$, $q > 2$; since $\mathfrak{so}(2,2) \cong \mathfrak{so}(2,1) \oplus \mathfrak{so}(2,1)$, $\mathfrak{so}(2,3) \cong \mathfrak{sp}(2,\mathbb{R})$, $\mathfrak{so}(2,4) \cong \mathfrak{su}(2,2)$, the first truly interesting case actually occurs for $q = 5$. Given a linear map

$$\alpha = [\mathbf{w}_1,\dots,\mathbf{w}_n]^\tau\colon \mathbb{R}^2 \longrightarrow \mathbb{R}^{2,q},$$

with the notation $\mathbf{w}_1 = (w_1^1, w_1^2), \mathbf{w}_2 = (w_2^1, w_2^2), \mathbf{w}_3 = (w_3^1, w_3^2), \dots, \mathbf{w}_n = (w_n^1, w_n^2)$ (where $(\cdot,\cdot)$ refers to a 2-component vector and *not* to the Lorentz form), the adjoint $\alpha^\dagger\colon \mathbb{R}^{2,q} \to \mathbb{R}^2$, or rather its negative, is given by the matrix

$$-\alpha^\dagger = \begin{bmatrix} w_1^2 & w_2^2 & -w_3^2 & \dots & -w_n^2 \\ -w_1^1 & -w_2^1 & w_3^1 & \dots & w_n^1 \end{bmatrix}.$$

Hence, with the notation $\mathbf{w}^1 = (w_1^1,\dots,w_n^1)$ etc.,

$$\mu_H(\alpha) = -\alpha^\dagger\alpha = \begin{bmatrix} (\mathbf{w}^1,\mathbf{w}^2) & (\mathbf{w}^2,\mathbf{w}^2) \\ -(\mathbf{w}^1,\mathbf{w}^1) & -(\mathbf{w}^1,\mathbf{w}^2) \end{bmatrix} \in \mathfrak{sp}(1,\mathbb{R}).$$

Under these circumstances, μ_H induces a continuous $\mathrm{SO}(2,q)$-equivariant bijection from W^{red} onto the closure of the $\mathrm{SO}(2,q)$-orbit $\mathcal{O}_{1,0} \cup \mathcal{O}_{0,1}$ which coincides with the subspace of square zero matrices in $\mathfrak{so}(2,q)$. As $\mathrm{SO}(2,q)^0$-orbits, $\mathcal{O}_{1,0}$ is holomorphic and $\mathcal{O}_{0,1}$ antiholomorphic. For completeness, we note that $e_1 = \mu_H(\alpha_1)$ and $e_2 = \mu_H(\alpha_2)$ where $\alpha_{1,2}\colon \mathbb{R}^2 \longrightarrow \mathbb{R}^n$ are given by

$$\alpha_1 = \begin{bmatrix} 1 & 0 & 1 & 0 & \dots & 0 \\ 0 & 1 & 0 & -1 & \dots & 0 \end{bmatrix}^\tau, \quad \alpha_2 = \begin{bmatrix} 0 & -1 & 1 & 0 & \dots & 0 \\ 1 & 0 & 0 & 1 & \dots & 0 \end{bmatrix}^\tau.$$

The remaining pseudoholomorphic nilpotent orbits $\mathcal{O}_{2,0}$, $\mathcal{O}_{0,2}$, and $\mathcal{O}_{1,1}$ of $\mathfrak{so}(2,q)$ are the connected components of the $\mathrm{Sp}(1,\mathbb{R})$-reduced space W_C, the space of closed $\mathrm{Sp}(1,\mathbb{R})$-orbits in the pre-image $\mu_H^{-1}(C)$ of the nilcone C in $\mathfrak{sp}(1,\mathbb{R})$. Indeed, the nilcone C is the disjoint union of the three nilpotent orbits $\mathcal{O}_0 = \{0\}, \mathcal{O}_0^+, \mathcal{O}_0^-$, the notation being that introduced in (3.2.2). Inspection shows that any matrix in C has square zero; this is also a very special case of Theorem 3.5.3 since $\mathcal{O}_0^+$ and $\mathcal{O}_0^-$ are pseudoholomorphic. Thus, given $\alpha \in \mu_H^{-1}(C)$,

$$(\mu_G(\alpha))^3 = (\alpha\alpha^\dagger)^3 = \alpha(\alpha^\dagger\alpha)^2\alpha^\dagger = \alpha(\mu_H(\alpha))^2\alpha^\dagger = 0$$

whence the momentum mapping $\mu_G\colon W \longrightarrow \mathfrak{so}(2,q)$ maps the pre-image $\mu_H^{-1}(C)$ into the space of matrices X in $\mathfrak{so}(2,q)$ with $X^3 = 0$. The classification of nilpotent orbits in $\mathfrak{so}(2,q)$ [**7**] implies that this map is onto the space of matrices X with $X^3 = 0$, in fact, induces a continuous bijection from W_C onto $\mathcal{O}_{2,0} \cup \mathcal{O}_{1,1} \cup \mathcal{O}_{0,2}$. A closer look reveals that

$$\mu_G(\mu_H^{-1}(\mathcal{O}_0^+)) = \mathcal{O}_{2,0} \cup \mathcal{O}_{0,2}, \quad \mu_G(\mu_H^{-1}(\mathcal{O}_0^-)) = \mathcal{O}_{1,1},$$

where the notation is that in (3.6). From this observation, it is not hard to deduce that the orbits $\mathcal{O}_{2,0}$, $\mathcal{O}_{0,2}$, and $\mathcal{O}_{1,1}$ are pseudoholomorphic. For example, since we may change α within its $\mathrm{Sp}(1,\mathbb{R})$-orbit without changing $\alpha\alpha^\dagger \in \mathfrak{so}(2,q)$, to study $\mathcal{O}_{2,0} \cup \mathcal{O}_{0,2}$, we may consider the smooth submanifold $\widetilde{\mathcal{O}_0^+}$ of $W = \mathbb{R}^{2n}$ which consists of those $\alpha = [\mathbf{w}_1, \mathbf{w}_2, \dots, \mathbf{w}_n]^\tau\colon \mathbb{R}^2 \longrightarrow \mathbb{R}^n$ which satisfy the equation

$$\mu_H(\alpha) = -\alpha^\dagger\alpha = \begin{bmatrix} (\mathbf{w}^1,\mathbf{w}^2) & (\mathbf{w}^2,\mathbf{w}^2) \\ -(\mathbf{w}^1,\mathbf{w}^1) & -(\mathbf{w}^1,\mathbf{w}^2) \end{bmatrix} = E = \begin{bmatrix} 0 & 1 \\ 0 & 0 \end{bmatrix}.$$

The stabilizer Z_E of $E \in \mathfrak{sp}(1,\mathbb{R}) = \mathfrak{sl}(2,\mathbb{R})$ is a one-dimensional unipotent subgroup in $\mathrm{Sp}(1,\mathbb{R})$; it acts freely on $\widetilde{\mathcal{O}_0^+}$ in such a way that the space of Z_E-orbits coincides with $\mathcal{O}_{2,0} \cup \mathcal{O}_{0,2}$. We claim that the $\mathrm{SO}(q,\mathbb{R})$-equivariant map

$$\pi\colon W = \mathbb{R}^{2n} \xrightarrow{\mu_G} \mathfrak{so}(2,q) = \mathfrak{g} = \mathfrak{k} \oplus \mathfrak{p} \xrightarrow{\mathrm{proj}} \mathfrak{p} = \mathbb{R}^{2q},$$

restricted to $\widetilde{\mathcal{O}_0^+}$, is a fibration onto its image in $\mathfrak{p} = \mathbb{R}^{2q}$ with fiber the real line. Indeed, by construction,

$$\pi[\mathbf{w}_1, \mathbf{w}_2, \dots, \mathbf{w}_n]^\tau = \begin{bmatrix} \mathbf{w}_3 \wedge \mathbf{w}_1 & \dots & \mathbf{w}_n \wedge \mathbf{w}_1 \\ \mathbf{w}_3 \wedge \mathbf{w}_2 & \dots & \mathbf{w}_n \wedge \mathbf{w}_2 \end{bmatrix}$$

Plainly, $\mathrm{SO}(q,\mathbb{R})$-equivariance reduces the study of π to that of the special case where $q = 3$ since, for any $\alpha = [\mathbf{w}_1, \mathbf{w}_2, \dots, \mathbf{w}_n]^\tau$, there is some $y \in \mathrm{SO}(q,\mathbb{R})$ such that

$$y[\mathbf{w}_1, \mathbf{w}_2, \mathbf{w}_3', \mathbf{w}_4', \mathbf{w}_5', 0, \dots, 0]^\tau = [\mathbf{w}_1, \mathbf{w}_2, \dots, \mathbf{w}_n]^\tau$$

for some $\mathbf{w}_3', \mathbf{w}_4', \mathbf{w}_5' \in \mathbb{R}^2$. However, $\mathfrak{so}(2,3) \cong \mathfrak{sp}(2,\mathbb{R})$ and, from the description of holomorphic nilpotent orbits in $\mathfrak{sp}(2,\mathbb{R})$ given in Section 3 above we know already that, for $q = 3$, the restriction of the map π to $\widetilde{\mathcal{O}_0^+}$ is a fibration of $\widetilde{\mathcal{O}_0^+}$ onto its image in $\mathfrak{p} = \mathbb{R}^6$ with fiber the real line. By $\mathrm{SO}(q,\mathbb{R})$-equivariance, this implies that, for general $q \geq 3$, the restriction of π to $\widetilde{\mathcal{O}_0^+}$ is a fibration of $\widetilde{\mathcal{O}_0^+}$ onto its image in $\mathfrak{p} = \mathbb{R}^{2q}$ with fiber the real line. Consequently the restriction to $\mathcal{O}_{2,0}$ of the projection from $\mathfrak{g} = \mathfrak{k} \oplus \mathfrak{p}$ to $\mathfrak{p}$ is a diffeomorphism onto its image, and the same is true of $\mathcal{O}_{0,2}$. The same kind of reasoning applies to $\mathcal{O}_{1,1}$. Thus the orbits $\mathcal{O}_{2,0}$, $\mathcal{O}_{1,1}$, and $\mathcal{O}_{0,2}$ are pseudoholomorphic. Henceforth we write $\mathcal{O}_1 = \mathcal{O}_{1,0}$ and $\mathcal{O}_2 = \mathcal{O}_{2,0}$.

6.2. THE COMPLEX STRUCTURE. The constituent $\mathfrak{p}$ in the Cartan decomposition $\mathfrak{so}(2,q) = \mathfrak{k} \oplus \mathfrak{p}$ consists of matrices in $\mathfrak{so}(2,q)$ of the kind $\begin{bmatrix} 0 & B \\ B^t & 0 \end{bmatrix}$ where B is an arbitrary real $(q \times 2)$-matrix and thus amounts to the space $\mathrm{M}_{q,2}(\mathbb{R})$ of real $(q \times 2)$-matrices. The requisite complex structure on $\mathrm{M}_{q,2}(\mathbb{R})$ is given by matrix multiplication

$$J \begin{bmatrix} x_1 & \dots & x_q \\ y_1 & \dots & y_q \end{bmatrix} = \begin{bmatrix} y_1 & \dots & y_q \\ -x_1 & \dots & -x_q \end{bmatrix}$$

whence complex analytically, $\overline{\mathcal{O}_2} \cong \mathfrak{p}^+$ amounts to a copy of $\mathbb{C}^q$. In this description,

$$e_1 = \mu_H(\alpha_1) = (i, 1, 0, \dots, 0) \in \mathbb{C}^q, \ e_2 = \mu_H(\alpha_2) = (1, i, 0, \dots, 0) \in \mathbb{C}^q,$$

and the action of the complexification $K^{\mathbb{C}} = \mathrm{SO}(2,\mathbb{C}) \times \mathrm{SO}(q,\mathbb{C})$ of the maximal compact subgroup $K = \mathrm{SO}(2,\mathbb{R}) \times \mathrm{SO}(q,\mathbb{R})$ of $\mathrm{SO}(2,q)$ is given by the ordinary $\mathrm{SO}(q,\mathbb{C})$-action on $\mathbb{C}^q$ together with the multiplication by non-zero complex numbers, when $\mathrm{SO}(2,\mathbb{C})$ is identified with $\mathbb{C}^*$. After introduction of the complex coordinates $w_1, \dots, w_q$ where $w_j = x_j + i y_j$ $(1 \leq j \leq q)$, under the homeomorphism from $\overline{\mathcal{O}_2}$ onto $\mathfrak{p}^+$, complex analytically, the closure $\overline{\mathcal{O}_1}$ is identified with the complex quadric in $\mathbb{C}^q$ given by the equation $f(w_1, \dots, w_q) = 0$ where

$$f(w_1, \dots, w_q) = w_1^2 + \dots + w_q^2.$$

In fact, $\mathcal{O}_1$ is generated by e_1, and the $K^{\mathbb{C}}$-orbit of $(i, 1, 0, \dots, 0) \in \mathbb{C}^q$ consists of the non-zero points of the quadric $w_1^2 + \dots + w_q^2 = 0$. The significance of the quadratic form f will be elucidated further in Theorem 7.1 below.

REMARK 6.3. The above observations combine to a proof of Theorem 3.6.3 and hence of statement (iii) of Theorem 3.6.2.

REMARK 6.4. Just as in the standard cases, even though as a complex analytic space, $\overline{\mathcal{O}_2}$ is affine and hence non-singular, as a stratified symplectic space and hence, as a normal Kähler space, it has *singularities*, the *singular locus* (in the sense of stratified Kähler spaces) being the complex quadric $\overline{\mathcal{O}_1}$.

7. Hermitian Jordan triple systems and pre-homogeneous spaces

In this section we will relate holomorphic nilpotent orbits with positive definite hermitian Jordan triple systems.

Recall that a pair $(\mathfrak{g}, \vartheta)$ consisting of a semisimple Lie algebra $\mathfrak{g}$ together with a 3-grading $\mathfrak{g} = \mathfrak{g}_{-1} + \mathfrak{g}_0 + \mathfrak{g}_1$ and an involution ϑ of $\mathfrak{g}$ such that $\vartheta\mathfrak{g}_j = \mathfrak{g}_{-j}$ $(j = -1, 0, 1)$ is called a *(non-degenerate) symmetric Lie algebra* ([**79**] I.7 p. 28) provided $\mathfrak{g}_0 = [\mathfrak{g}_{-1}, \mathfrak{g}_1]$. Given a symmetric Lie algebra $(\mathfrak{g}, \vartheta)$, the abelian subalgebra $V = \mathfrak{g}_{-1}$ (as well as the abelian subalgebra $\mathfrak{g}_1$), endowed with the trilinear operation $\{\cdot, \cdot, \cdot\}$ defined by

$$\{x, y, z\} = -\frac{1}{2}[[x, \vartheta y], z]$$

is what is called a *non-degenerate J(ordan) T(riple) S(ystem)*—the defining axioms are not reproduced here, cf. [**79**] p. 22—and every non-degenerate JTS arises in this fashion, see [**79**] p. 28 and the references there. For a simple Lie algebra $(\mathfrak{g}, z)$ of hermitian type, the decomposition (3.3.4(r)) (where r is the real rank of $\mathfrak{g}$) yields a real symmetric Lie algebra if and only if $(\mathfrak{g}, z)$ is regular, the requisite involution being the Cartan involution.

A (non-degenerate) real JTS $(V, \{\cdot, \cdot, \cdot\})$, together with a complex structure (on V) such that $\{\cdot, \cdot, \cdot\}$ is $\mathbb{C}$-linear in the first and third variable and $\mathbb{C}$-antilinear in the second one is said to be a *hermitian* JTS, and it is said to be *positive definite* provided the hermitian form τ' on V ($\mathbb{C}$-linear in the second variable) given by the trace $\tau'(x, y)$ (taken over $\mathbb{C}$) of the $\mathbb{C}$-linear transformation $y\Box x$ of V given by $(y\Box x)z = \{y, x, z\}$ is positive definite $(x, y, z \in V)$.

Let $(\mathfrak{g}, z)$ be a simple Lie algebra of hermitian type, with Cartan decomposition $\mathfrak{g} = \mathfrak{k} \oplus \mathfrak{p}$. Its complexification $\mathfrak{g}^{\mathbb{C}} = \mathfrak{p}^+ \oplus \mathfrak{k}^{\mathbb{C}} \oplus \mathfrak{p}^-$ inherits a real symmetric Lie algebra structure, the requisite involution being complex conjugation (but, beware, this is *not* the symmetric Lie algebra structure mentioned earlier). Thus $(\mathfrak{g}, z)$ determines the (non-degenerate) JTS $(V, \{\cdot, \cdot, \cdot\})$ where $V = \mathfrak{p}^+$. It is known that this JTS is hermitian and positive definite and that every positive definite hermitian JTS arises in this fashion, cf. [**79**] (Proposition II.3.3 on p. 56); indeed, with the appropriate notions of morphism, associating with a (real) semisimple Lie algebra of hermitian type $(\mathfrak{g}, z)$ the JTS $(\mathfrak{p}^+, \{\cdot, \cdot, \cdot\})$ yields an equivalence of categories between (real) semisimple Lie algebras of hermitian type and positive definite hermitian JTS's [**79**] (Proposition II.8.2 on p. 85). Under this correspondence, the regular semisimple Lie algebras of hermitian type correspond to positive definite hermitian JTS's which "arise" from a Jordan algebra (with unit element), in the following sense: A finite dimensional (non-associative) algebra U satisfying $xy = yx$ and $x^2(xy) = x(x^2y)$ $(x, y \in U)$ is called a *Jordan algebra*. Any Jordan algebra U becomes a JTS when the triple product $\{\cdot, \cdot, \cdot\}$ is defined by

$$\{x, y, z\} = (xy)z + x(yz) - y(xz), \quad x, y, z \in U,$$

and we then say that the JTS *arises* from the Jordan algebra U. When U has a unit element e, the Jordan product is determined by the triple product by means of

$$\{x, y, e\} (= \{x, e, y\} = \{e, x, y\}) = xy, \quad x, y \in U.$$

A representation E of a complex reductive Lie group is said to be a *pre-homogeneous space* [**90**] provided it has a Zariski-open orbit, referred to henceforth as the *top* orbit, and it is said to be *regular* [**80**] provided the complement of the top orbit is a hypersurface or, equivalently, provided the isotropy subgroups of the points in the top orbit are reductive (cf. p. 96 of [**68**]). For an irreducible regular pre-homogeneous space, where "irreducible" means that the representation is irreducible, the hypersurface arising as the complement of the top orbit is known to be irreducible; the polynomial which it defines is a *relative invariant* (i. e. it is a "twisted" invariant, with reference to a suitable character) of the representation (p. 96 of [**68**]) which is unique up to a constant and referred to as the *fundamental relative invariant*; this fundamental relative invariant then generates the relative invariants.

Given a simple Lie algebra $(\mathfrak{g}, z)$ of hermitian type, the space $\mathfrak{p}^+$ is plainly a pre-homogeneous space for the complexification $K^{\mathbb{C}}$ of the compact constituent K in the Cartan decomposition of the adjoint group G of $\mathfrak{g}$. These pre-homogeneous spaces are precisely those which arise from positive definite hermitian JTS's. In [**68**], they have been classified in terms of root systems (without reference to the notion of Lie algebra of hermitian type). For intelligibility, we reproduce explicit descriptions of the $K^{\mathbb{C}}$-actions on $\mathfrak{p}^+$ in the classical cases and recall which ones are regular:

$\mathfrak{g} = \mathfrak{sp}(\ell, \mathbb{R})$: $\mathfrak{p}^+ = \mathrm{S}^2_{\mathbb{C}}[\mathbb{C}^\ell]$ is regular; $K^{\mathbb{C}} = \mathrm{GL}(\ell, \mathbb{C})$, the action on $\mathrm{S}^2_{\mathbb{C}}[\mathbb{C}^\ell]$ being given by

$$x \cdot S = xSx^t, \quad x \in \mathrm{GL}(\ell, \mathbb{C}),\ S \in \mathrm{S}^2_{\mathbb{C}}[\mathbb{C}^\ell].$$

This is the second symmetric power of the standard $\mathrm{GL}(\ell, \mathbb{C})$-representation.

$\mathfrak{g} = \mathfrak{u}(p, q)$: $\mathfrak{p}^+ = \mathrm{M}_{q,p}(\mathbb{C})$ is regular if and only if $p = q$; $K^{\mathbb{C}} = \mathrm{GL}(p, \mathbb{C}) \times \mathrm{GL}(q, \mathbb{C})$, the action on $\mathrm{M}_{q,p}(\mathbb{C})$ being given by

$$(x, y) \cdot M = xMy^t, \quad x \in \mathrm{GL}(p, \mathbb{C}),\ y \in \mathrm{GL}(q, \mathbb{C}),\ M \in \mathrm{M}_{q,p}(\mathbb{C});$$

$\mathfrak{g} = \mathfrak{so}^*(2n)$: $\mathfrak{p}^+ = \Lambda^2[\mathbb{C}^n]$ is regular if and only if n is even; $K^{\mathbb{C}} = \mathrm{GL}(n, \mathbb{C})$, the action on $\Lambda^2[\mathbb{C}^n]$ being given by

$$x \cdot A = xAx^t, \quad x \in \mathrm{GL}(n, \mathbb{C}),\ A \in \Lambda^2[\mathbb{C}^n].$$

This is the second exterior power of the standard $\mathrm{GL}(n, \mathbb{C})$-representation.

$\mathfrak{g} = \mathfrak{so}(2, q)$: $\mathfrak{p}^+ = \mathbb{C}^q$ is regular; $K^{\mathbb{C}} = \mathrm{SO}(2, \mathbb{C}) \times \mathrm{SO}(q, \mathbb{C})$, the action on $\mathbb{C}^q$ being given by the ordinary $\mathrm{SO}(q, \mathbb{C})$-action on $\mathbb{C}^q$ together with the multiplication by non-zero complex numbers, when $\mathrm{SO}(2, \mathbb{C})$ is identified with $\mathbb{C}^*$.

Thus, $\mathfrak{p}^+$ is regular if and only if $(\mathfrak{g}, z)$ is regular (cf. (3.3) above), that is, if and only if its relative root system is of type (C_r) (where r is the real rank). Furthermore, the property of being regular is equivalent to the corresponding positive definite hermitian JTS to arise from a formally real Jordan algebra, cf. [**79**] I.8 p. 31 (also referred to as *euclidean* Jordan algebra in the literature [**14**]), and the complexification of these formally real Jordan algebras gives the simple complex Jordan algebra structure on the constituent $\mathfrak{p}^+$. Plainly, in these cases, whenever $\mathfrak{p}^+$ is regular, it is irreducible. Recall that the *rank* of a non-zero element of a euclidean as well as of a positive definite hermitian Jordan algebra, referred to henceforth as its *Jordan rank*, is the number of non-zero eigenvalues in its spectral decomposition, with multiplicities counted [**14**] (p. 77).

THEOREM 7.1. *In the regular classical cases, under the homeomorphism (induced by the projection from $\mathfrak{g}$ to $\mathfrak{p}$) from the closure $\overline{\mathcal{O}}$ of the principal holomorphic nilpotent orbit $\mathcal{O}$ onto the complex vector space $\mathfrak{p}^+$, the G-orbit stratification of $\overline{\mathcal{O}}$*

passes to the stratification of $\mathfrak{p}^+$ *by Jordan rank, with reference to the complex simple Jordan algebra structure on* $\mathfrak{p}^+$*, and that stratification, in turn, coincides with the* $K^{\mathbb{C}}$*-orbit stratification of* $\mathfrak{p}^+$*. In the standard cases, when* $\mathfrak{p}^+$ *is realized as a space of complex matrices as indicated above, the fundamental relative invariant (Bernstein-Sato polynomial) is the determinant. For* $\mathfrak{g} = \mathfrak{so}(2,q)$*, in suitable complex coordinates* $w_1, \ldots, w_q$ *on* $\overline{\mathcal{O}_2} \cong \mathfrak{p}^+ \cong \mathbb{C}^q$ *(e.g. those introduced in* (6.2)*), the fundamental relative invariant (Bernstein-Sato polynomial)* f *is the quadratic form*

$$f(w_1, \ldots, w_q) = w_1^2 + \cdots + w_q^2.$$

Thus the stratified symplectic Poisson structure detects the stratification by Jordan rank.

PROOF. The statement involving the Jordan rank is just a reformulation of the description of the strata in Theorems 3.5.5 and 3.6.3. To see that the fundamental relative invariants have the asserted form, we note that, by Theorem 5.4, in each of the cases $(\mathfrak{g}, \mathfrak{p}^+, r) = (\mathfrak{sp}(\ell, \mathbb{R}), \mathfrak{p}^+ = \mathrm{S}^2_{\mathbb{C}}[\mathbb{C}^\ell], \ell)$, $(\mathfrak{g}, \mathfrak{p}^+, r) = (\mathfrak{su}(p,p), \mathrm{M}_{p,p}(\mathbb{C}), p)$, $(\mathfrak{g}, \mathfrak{p}^+, r) = (\mathfrak{so}^*(4\ell), \Lambda^2[\mathbb{C}^{2\ell}], \ell)$, the complement of the top stratum $\mathcal{O}_r$ is the hypersurface $\overline{\mathcal{O}_{r-1}}$ which is given by a single determinantal equation. The determinant is clearly a relative invariant. For $\mathfrak{g} = \mathfrak{so}(2,q)$, in the coordinates $w_1, \ldots, w_q$ on $\overline{\mathcal{O}_2} \cong \mathfrak{p}^+ \cong \mathbb{C}^q$ introduced in (6.2), the fundamental relative invariant f on $\overline{\mathcal{O}_2} \cong \mathfrak{p}^+ \cong \mathbb{C}^q$ has the form

$$f(w_1, \ldots, w_q) = w_1^2 + \cdots + w_q^2.$$

This is truly a relative $K^{\mathbb{C}}$-invariant, the requisite character for $K^{\mathbb{C}}$ being given by the projection to $\mathrm{SO}(2,\mathbb{C}) \cong \mathbb{C}^*$, followed by the squaring map from $\mathbb{C}^*$ to itself. Finally, it is a standard fact that a stratified symplectic Poisson algebra detects the strata. □

In the regular standard cases, the fundamental relative invariants for the complexification of the compact constituent in the Cartan decomposition of G therefore arise *from the relations among the* K*-invariants describing the complex analytic model* $W_J//K^{\mathbb{C}}$ of $\overline{\mathcal{O}_{r-1}}$ since these yield the defining equations for $\overline{\mathcal{O}_r}$ as a complex affine variety. Thus the fundamental relative invariants are seen as resulting from an application of the "Second Main Theorem" of Invariant Theory.

REMARK 7.2. See **[14, 65, 79]** for notation and details. For $\mathbb{K} = \mathbb{R}$, $\mathbb{C}$, $\mathbb{H}$, consider the *real special Jordan algebra* of hermitian matrices $\mathcal{H}_n(\mathbb{K})$ over $\mathbb{K}$ ($\mathcal{H}_n(\mathbb{R}) = S^2_{\mathbb{R}}[\mathbb{R}^n]$), with Jordan product $\circ$ given by $x \circ y = \frac{1}{2}(xy + yx)$ $(x, y \in \mathcal{H}_n(\mathbb{K}))$. The *Kantor-Koecher-Tits* construction of superstructure algebra for $\mathcal{H}_n(\mathbb{K})$ over $\mathbb{K}$ yields the regular standard simple Lie algebras of hermitian type, together with the appropriate real symmetric Lie algebra decompositions (3.3.4(r)) (where r is the real rank), as follows:

$$\begin{aligned}
\mathbb{K} = \mathbb{R}&: \quad \mathcal{H}_n(\mathbb{R}) \oplus \mathfrak{gl}(n,\mathbb{R}) \oplus \mathcal{H}_n(\mathbb{R}) &&= \mathfrak{sp}(n,\mathbb{R})\\
\mathbb{K} = \mathbb{C}&: \quad \mathcal{H}_n(\mathbb{C}) \oplus \mathfrak{gl}(n,\mathbb{C}) \oplus \mathcal{H}_n(\mathbb{C}) &&= \mathfrak{u}(n,n)\\
\mathbb{K} = \mathbb{H}&: \quad \mathcal{H}_n(\mathbb{H}) \oplus \mathfrak{gl}(n,\mathbb{H}) \oplus \mathcal{H}_n(\mathbb{H}) &&= \mathfrak{so}^*(4n)
\end{aligned}$$

Likewise, the superstructure algebra construction for the euclidean Jordan algebra $J(1, q-1)$ (arising from the Lorentz form of type $(1, q-1)$ on $\mathbb{R}^q$) yields the

Lie algebra $\mathfrak{so}(2,q)$, together with the corresponding real symmetric Lie algebra decomposition (3.3.4(2)), which has the form

$$J(1,q-1) \oplus \mathbb{R}h^2 \oplus \mathfrak{so}(1,q-1) \oplus J(1,q-1) = \mathfrak{so}(2,q).$$

We will now show that, in the non-regular classical cases, the stratification can still be described by Jordan rank: The notion of rank of an element of a positive definite hermitian JTS (as well as of a real positive definite JTS) has been introduced in [**62**] (11.6), see also [**72**] (where a general notion of rank, written r, is introduced without a name on p. 2625); accordingly we will continue to refer to the *Jordan rank* of such an element. For $\mathfrak{g} = \mathfrak{su}(p,q)$, when $p > q = r$ (so that $(\mathfrak{g}, z)$ is non-regular), the projection from $\mathfrak{g}$ to $\mathfrak{p}$, followed by the isomorphism from $\mathfrak{p}$ onto $\mathfrak{p}^+ = \mathrm{M}_{q,p}(\mathbb{C})$, identifies the closure $\overline{\mathcal{O}_q}$ of the principal holomorphic nilpotent orbit $\mathcal{O}_q$ with the complex vector space underlying the positive definite hermitian Jordan triple system $\mathrm{M}_{q,p}(\mathbb{C})$ (the triple product being given by $\{x,y,z\} = x\overline{y}^t z + z\overline{y}^t x$). In this fashion, for $1 \leq s < r$, $\mathcal{O}_s$ amounts the space of complex $(q \times p)$-matrices of rank s and $\overline{\mathcal{O}_s}$ with that of complex $(q \times p)$-matrices of rank at most s, and the ordinary rank of a matrix coincides with its Jordan rank. Likewise, in case $\mathfrak{g} = \mathfrak{so}^*(2n)$, when $n = 2\ell + 1$ (so that $(\mathfrak{g}, z)$ is non-regular), the corresponding map from $\mathfrak{g}$ to $\mathfrak{p}^+ = \Lambda^2[\mathbb{C}^n]$ identifies the closure $\overline{\mathcal{O}_\ell}$ of the principal holomorphic nilpotent orbit $\mathcal{O}_\ell$ with the complex vector space underlying the positive definite hermitian Jordan triple system $\Lambda^2[\mathbb{C}^n]$. For $1 \leq s \leq r = \ell$, $\mathcal{O}_s$ thus amounts to the space of skew-symmetric complex $(n \times n)$-matrices of rank $2s$ and $\overline{\mathcal{O}_s}$ to that of skew-symmetric complex $(n \times n)$-matrices of rank at most $2s$, and the ordinary rank of a matrix is twice its Jordan rank. Summarizing, we spell out the following.

THEOREM 7.3. *In the classical cases, under the homeomorphism from the closure $\overline{\mathcal{O}}$ of the principal holomorphic nilpotent orbit $\mathcal{O}$ onto the complex vector space $\mathfrak{p}^+$, the G-orbit stratification of $\overline{\mathcal{O}}$ passes to the stratification of $\mathfrak{p}^+$ by Jordan rank with reference to the (positive definite hermitian) JTS-structure on $\mathfrak{p}^+$, and that stratification, in turn, coincides with the $K^{\mathbb{C}}$-orbit stratification of $\mathfrak{p}^+$. Thus the induced stratified symplectic Poisson structure on $\mathfrak{p}^+$ turns the latter into a normal Kähler space and detects the stratification by Jordan rank.* □

All geometric information about holomorphic nilpotent orbits in a simple classical Lie algebra of hermitian type is therefore encoded in the corresponding JTS. Theorem 7.3 entails in particular that the pre-homogeneous spaces arising from positive definite hermitian JTS's and classified in [**68**] carry much more structure than that made explicit in that reference.

8. The exceptional cases

In this section we recollect some information relevant for the exceptional cases; see e. g. [**14**], [**47**], [**77**], and [**79**] (III.4 p. 117/118) for details and notation. We then describe the geometry of the corresponding holomorphic nilpotent orbits. Until the end of this section, $(\mathfrak{g}, z)$ will be one of the two simple exceptional Lie algebras of hermitian type; as before, we write $\mathfrak{g} = \mathfrak{k} \oplus \mathfrak{p}$ for the Cartan decomposition and $\mathfrak{g}_{\mathbb{C}} = \mathfrak{p}^- \oplus \mathfrak{k}_{\mathbb{C}} \oplus \mathfrak{p}^+$ for the complexification thereof; further, we denote the real rank by r, the adjoint group of $\mathfrak{g}$ by G, and the maximal compact subgroup determined by the choice of z by K.

8.1. Octonions and Albert algebras. We shall use the neutral notation $\mathbb{K}$ for $\mathbb{R}$ and/or $\mathbb{C}$ if need be. Let $\mathbb{O} = \mathbb{O}_{\mathbb{R}}$ be the real *octonion algebra of Cayley numbers*, also called *Cayley division algebra*, and write $\mathbb{O}_{\mathbb{C}}$ for its complexification; this is a complex octonion algebra. As usual, for $a \in \mathbb{O}$ as well as for $a \in \mathbb{O}_{\mathbb{C}}$, we write $a \mapsto \overline{a}$ for the operation of conjugation, and the *norm* $n(a) \in \mathbb{K}$ and *trace* $t(a) \in \mathbb{K}$ are defined by $n(a) = a\overline{a}$ and $t(a) = a + \overline{a}$ where $\mathbb{K}$ is viewed as a subalgebra of $\mathbb{O}_{\mathbb{K}}$ in the obvious fashion. For $m = 2$ and $m = 3$, let $\mathcal{H}_m(\mathbb{O})$ be the euclidean Jordan algebra of positive definite hermitian $(m \times m)$-matrices over $\mathbb{O}$. The Jordan algebra $\mathcal{H}_2(\mathbb{O})$ has rank 2 and is isomorphic to $J(1,9)$—in view of the classification, every rank two hermitian Jordan algebra arises from a quadratic form in this fashion—and $\mathcal{H}_3(\mathbb{O})$ is the real reduced exceptional central simple rank 3 Jordan algebra of dimension 27, a real *Albert algebra*. The complexification of the latter is the (reduced central simple) rank 3 *complex exceptional Jordan algebra* $\mathcal{H}_3(\mathbb{O}_{\mathbb{C}})$, a complex *Albert algebra*; cf. e. g. VIII.5 on p. 159 of [**14**], p. 121 of [**47**]. Here the property of being *reduced* amounts to the existence of three non-zero orthogonal idempotents, and an Albert algebra being *exceptional* means that it cannot be embedded into any associative algebra whatsoever. Whenever we wish to treat the two Jordan algebras simultaneously, we write $\mathcal{H}_3(\mathbb{O}_{\mathbb{K}})$, for $\mathbb{K} = \mathbb{R}$ and $\mathbb{K} = \mathbb{C}$; in [**47**] (p. 17), $\mathcal{H}_3(\mathbb{O}_{\mathbb{K}})$ is denoted by $\mathrm{H}(\mathcal{C}_3, \gamma)$ with hermitian form $\gamma = \mathrm{diag}(1,1,1)$ where $\mathcal{C}$ refers to the Cayley division algebra over $\mathbb{R}$ as well as to the complexification thereof. Thus the Jordan algebra $\mathcal{H}_3(\mathbb{O}_{\mathbb{K}})$ consists of (3×3)-matrices

$$A = \begin{bmatrix} \alpha_1 & a_3 & \overline{a_2} \\ \overline{a_3} & \alpha_2 & a_1 \\ a_2 & \overline{a_1} & \alpha_3 \end{bmatrix}$$

where $\alpha_1, \alpha_2, \alpha_3 \in \mathbb{K}$ and $a_1, a_2, a_3 \in \mathbb{O}_{\mathbb{K}}$, and the Jordan product $\circ$ is given by $x \circ y = \frac{1}{2}(xy + yx)$ $(x, y \in \mathcal{H}_3(\mathbb{O}_{\mathbb{K}}))$. Such a matrix A has *generic norm* $\nu(A)$ given by

$$\nu(A) = \alpha_1\alpha_2\alpha_3 + t(a_3a_1a_2) - \alpha_1 n(a_1) - \alpha_2 n(a_2) - \alpha_3 n(a_3).$$

The generic norm ν amounts to Freudenthal's "formally defined determinant" [**15**].

8.2. $\mathfrak{g} = \mathfrak{e}_{7(-25)}$. This Lie algebra arises from the superstructure algebra construction (cf. e. g. [**65**]) for the positive definite hermitian Jordan algebra $\mathcal{H}_3(\mathbb{O})$, cf. [**47**] (No. 3 p. 121), [**65**]. The real rank r of $\mathfrak{e}_{7(-25)}$ equals 3, the maximal compact subgroup K of the adjoint group is locally isomorphic to $E_{6(-78)} \times \mathrm{SO}(2,\mathbb{R})$ ($\mathrm{E}_{6(-78)}$ being a compact form), $\dim \mathfrak{k} = 79$, and $\dim \mathfrak{p} = 54$.

The stratification of the closure of the principal holomorphic nilpotent orbit may be described in terms of the corresponding decompositions (3.3.4(k)) ($1 \leq k \leq 3$); in particular, $\dim \mathfrak{z}_{\mathfrak{g}}(e^3) = 52 + 27 = 79 = \dim \mathfrak{k}$ whence $\dim \mathcal{O}_3$ equals $\dim \mathfrak{p}$ ($\mathcal{O}_3 = Ge^3$). Indeed, direct inspection confirms that the projection from the closure

$\overline{\mathcal{O}_3}$ of the principal holomorphic nilpotent orbit $\mathcal{O}_3$ to $\mathfrak{p}$ is a homeomorphism *onto* $\mathfrak{p}$, as claimed in Theorem 3.3.11.

8.3. $\mathfrak{g} = \mathfrak{e}_{6(-14)}$: With respect to the primitive idempotent $e_{3,3}$, the *Peirce decomposition* (cf. e. g. [**79**] V.6 p. 242 ff.) of $\mathcal{H}_3(\mathbb{O})$, viewed as a JTS, has the form

$$\mathcal{H}_3(\mathbb{O}) = V_0 \oplus V_{\frac{1}{2}} \oplus V_1;$$

here V_0 is the rank 2 Jordan algebra which is a copy of $\mathcal{H}_2(\mathbb{O})$, V_1 is the rank 1 Jordan algebra which is the (real) span of $e_{3,3}$, and $V_{\frac{1}{2}}$ is the JTS of (3×3)-matrices $A = \begin{bmatrix} 0 & 0 & \overline{a_2} \\ 0 & 0 & a_1 \\ a_2 & \overline{a_1} & 0 \end{bmatrix}$ where $a_1, a_2 \in \mathbb{O}$. The triple product on $\mathcal{H}_3(\mathbb{O})$ induces a triple product on $V_{\frac{1}{2}}$ turning the latter into a positive definite real JTS of dimension 16. We identify the elements of $V_{\frac{1}{2}}$ with the space $\mathrm{M}_{1,2}(\mathbb{O})$ of (1×2)-matrices over $\mathbb{O}$ by means of the assignment to $[a_1, a_2]^\tau$ of $\begin{bmatrix} 0 & 0 & \overline{a_2} \\ 0 & 0 & a_1 \\ a_2 & \overline{a_1} & 0 \end{bmatrix}$. The superstructure algebra construction for this positive definite real JTS yields the exceptional Lie algebra $\mathfrak{e}_{6(-14)}$. The real rank r of $\mathfrak{e}_{6(-14)}$ equals 2, the maximal compact subgroup K of the adjoint group G is isomorphic to $\mathrm{Spin}_{\mathrm{c}}(10,\mathbb{R})$, and $\dim \mathfrak{k} = 46$, $\dim \mathfrak{p} = 32$. To identify the representation of $\mathrm{Spin}_{\mathrm{c}}(10,\mathbb{R})$ on $\mathfrak{p}$ we note that, $\mathfrak{p}$ being endowed with the complex structure J_z, the restriction of this representation to $\mathrm{Spin}(10,\mathbb{R}) \subseteq \mathrm{Spin}_{\mathrm{c}}(10,\mathbb{R})$ is the positive half-spin representation $\mathcal{D}^5_+$ of complex dimension 16; here $\mathcal{D}^5_+$ is one of the two irreducible modules for the complex Clifford algebra C_9.

As pointed out earlier, the stratification of the closure of the principal holomorphic nilpotent orbit may be described in terms of the corresponding decompositions (3.3.4(k)) $(1 \leq k \leq 2)$; in particular, $\dim \mathfrak{z}_{\mathfrak{g}}(e^2) = \dim \mathfrak{k}$, whence $\dim \mathcal{O}_2 = \dim \mathfrak{p}$ $(\mathcal{O}_2 = Ge^2)$. Again, direct inspection confirms that the projection from the closure $\overline{\mathcal{O}_2}$ of the principal holomorphic nilpotent orbit $\mathcal{O}_2$ to $\mathfrak{p}$ is a homeomorphism *onto* $\mathfrak{p}$, as claimed in Theorem 3.3.11.

8.4. THE STATEMENT OF THEOREM 7.3 FOR THE EXCEPTIONAL CASES. The generic norm ν for $\mathcal{H}_3(\mathbb{O}_{\mathbb{C}})$ is a homogeneous complex cubic form. The notion of *Jordan rank* has been recalled in Section 7 above. For intelligibility, we note that $\mathfrak{g}_{\mathbb{C}} = \mathfrak{e}_7(\mathbb{C})$ is the superstructure algebra of $\mathcal{H}_3(\mathbb{O}_{\mathbb{C}})$, and that the structure group $K^{\mathbb{C}}$ of $\mathcal{H}_3(\mathbb{O}_{\mathbb{C}})$ is locally isomorphic to $\mathbb{C}^* \times E_6(\mathbb{C})$.

THEOREM 8.4.1. *For $\mathfrak{g} = \mathfrak{e}_{7(-25)}$, the pre-homogeneous space representation on $\mathfrak{p}^+$ is the resulting $K^{\mathbb{C}}$-representation on $\mathcal{H}_3(\mathbb{O}_{\mathbb{C}})$, and the relative invariant (Bernstein-Sato polynomial) is the generic norm ν. Furthermore, under the projection from $\mathfrak{g} = \mathfrak{k} \oplus \mathfrak{p}$ to $\mathfrak{p}$, followed by the canonical complex linear isomorphism from $\mathfrak{p}$ to $\mathfrak{p}^+$, the closure $\overline{\mathcal{O}_3}$ of the top holomorphic nilpotent orbit is identified with $\mathfrak{p}^+ = \mathcal{H}_3(\mathbb{O}_{\mathbb{C}})$ in such a way that, for $0 \leq s \leq 3$, $\mathcal{O}_s$ is mapped diffeomorphically onto the smooth complex affine variety which consists of Jordan rank s elements whence, for $1 \leq s < 3$, as a complex analytic space, $\overline{\mathcal{O}_s}$ amounts to the complex affine variety in $\mathfrak{p}^+$ which consists of elements in $\mathfrak{p}^+$ which have Jordan rank at most s. Thus requiring that the generic norm vanish yields an explicit equation defining $\overline{\mathcal{O}_2}$ as an affine complex variety in $\mathfrak{p}^+ = \mathcal{H}_3(\mathbb{O}_{\mathbb{C}})$ $(\cong \mathbb{C}^{27})$. As affine complex varieties, $\overline{\mathcal{O}_1}$ and $\overline{\mathcal{O}_2}$ are normal (und so is $\overline{\mathcal{O}_3}$, being an affine space).*

The top stratum $\mathcal{O}_3$ is thus identified with the Jordan-invertible elements in $\mathcal{H}_3(\mathbb{O}_\mathbb{C})$. In the literature, the generic norm hypersurface $\overline{\mathcal{O}_2} = \nu^{-1}(0) \subseteq \mathcal{H}_3(\mathbb{O}_\mathbb{C})$ is sometimes referred to as *Freudenthal's cubic*; we write it as $\mathcal{F}$. We note that, for $1 \leq s \leq 3$, the group $E_6(\mathbb{C})$ ($\subseteq K^\mathbb{C}$) acts transitively on (the image of) each orbit $\mathcal{O}_s$ (in $\mathfrak{p}^+$), and that, in view of the infinitesimal structure given in Table XIII in [**12**] (p. 516), the point stabilizers are the following subgroups of $E_6(\mathbb{C})$:
$F_4(\mathbb{C})$ (for $\mathcal{O}_3$), $\mathrm{Spin}(9,\mathbb{C}) \cdot (\mathbb{C}^*)^{16}$ (for $\mathcal{O}_2$), $\mathrm{Spin}(10,\mathbb{C}) \cdot (\mathbb{C}^*)^{16}$ (for $\mathcal{O}_1$).

PROOF OF THEOREM 8.4.1. The first statement is a special case of Theorem 3.3.11. Since $\mathcal{F}$ is an irreducible complex affine hypersurface which is non-singular in codimension 1, in view of standard algebraic geometry results, $\overline{\mathcal{O}_2} = \mathcal{F}$ is normal. The normality of the lowest non-trivial stratum $\overline{\mathcal{O}_1}$ will be justified in (10.7) below. □

Let V be a complex vector space of even (complex) dimension n, endowed with a non-degenerate quadratic form q, let $C(V,q)$ be the corresponding complex Clifford algebra, and let $\mathcal{S}^\mathbb{C}$ be an irreducible complex spinor space, i. e. a fundamental Clifford module. Recall that a spinor $\sigma \in \mathcal{S}^\mathbb{C}$ is said to be *pure* provided the kernel of the $\mathbb{C}$-linear map

$$j_\sigma \colon V \longrightarrow \mathcal{S}^\mathbb{C}, \quad j_\sigma(v) = v \cdot \sigma \quad \text{(Clifford multiplication)}$$

has dimension $\frac{n}{2}$; see e. g. [**59**] (IV.9 p. 336) for details. For $n = 10$, $\mathcal{S}^\mathbb{C} = \mathcal{D}^5$ and, as a $\mathrm{Spin}(10,\mathbb{R})$-representation, $\mathcal{S}^\mathbb{C}$ decomposes into the positive and negative complex half-spin representations $\mathcal{S}^\mathbb{C}_+ = \mathcal{D}^5_+$ and $\mathcal{S}^\mathbb{C}_- = \mathcal{D}^5_-$.

THEOREM 8.4.2. *For* $\mathfrak{g} = \mathfrak{e}_{6(-14)}$, *under the projection from* $\mathfrak{g} = \mathfrak{k} \oplus \mathfrak{p}$ *to* $\mathfrak{p}$, *followed by the canonical complex linear isomorphism from* $\mathfrak{p}$ *to* $\mathfrak{p}^+$, *the orbit* $\mathcal{O}_1$ *is identified with the space of positive pure spinors in* $\mathfrak{p}^+ \cong \mathcal{S}^\mathbb{C}_+$. *As a complex analytic space,* $\overline{\mathcal{O}_1}$ *amounts to the complex affine variety in* $\mathfrak{p}^+$ *which consists of elements in* $\mathfrak{p}^+$ *which have Jordan rank at most* 1. *As an affine complex variety,* $\overline{\mathcal{O}_1}$ *is normal (and so is* $\overline{\mathcal{O}_2}$, *being an affine space).*

We note that the group $\mathrm{Spin}(10,\mathbb{C})$ ($\subseteq K^\mathbb{C}$) acts transitively on the images of $\mathcal{O}_1$ and $\mathcal{O}_2$ in $\mathfrak{p}^+$, and that, in view of Proposition 2 in [**45**] (p. 1011), the point stabilizers for the orbits $\mathcal{O}_2$ and $\mathcal{O}_1$ are the subgroups $\mathrm{Spin}(7,\mathbb{C}) \cdot (\mathbb{C}^*)^8$ and $\mathrm{SL}(5,\mathbb{C}) \cdot (\mathbb{C}^*)^{10}$, respectively, of $\mathrm{Spin}(10,\mathbb{C})$.

PROOF. The first statement is a special case of Theorem 3.3.11. Since the representation $\mathcal{S}^\mathbb{C}_+$ has only two non-zero $\mathrm{Spin}(10,\mathbb{C})$-orbits, the orbit $\mathcal{O}_1$ is necessarily that of pure spinors. The normality of the stratum $\overline{\mathcal{O}_1}$ will be justified in (10.7) below. □

9. Contraction of semisimple holomorphic orbits

The idea of contraction (as the operation inverse to a deformation) goes back to [**46**]; a leisurely introduction into the notion of contraction we will use may be found in Chap. V of [**18**]. Let $(\mathfrak{g}, z)$ be a simple Lie algebra of Hermitian type and, as before, write G for its adjoint group and K for the maximal compact subgroup of G determined by the choice z of H-element (i. e. $\mathrm{Lie}(K) = \mathfrak{k} = \ker(\mathrm{ad}_z)$). Thus K is the stabilizer $Z_G(z)$ of z, and the Lie algebra $\mathfrak{k}$ of K equals the stabilizer algebra $\mathfrak{z}_{\mathfrak{g}}(z)$. With reference to the decomposition $\mathfrak{g} = \mathfrak{n}_r^- \oplus \mathfrak{l}_r \oplus \mathfrak{n}_r^+$, cf. (3.3.4($r$)) where, as before, r is the real rank of $\mathfrak{g}$, let $\mathfrak{k}_0 = \mathfrak{k} \cap \mathfrak{l}_r$, and consider the filtered Lie algebra $\mathfrak{k}_0 \subseteq \mathfrak{k}$; its associated graded Lie algebra $E^0(\mathfrak{k}) = \mathfrak{k}_0 \ltimes \mathfrak{n}_r^+$ is the semi-direct product of its compact constituent $\mathfrak{k}_0$ with the nilpotent Lie algebra $\mathfrak{n}_r^+$, the $\mathfrak{k}_0$-action on $\mathfrak{n}_r^+$ being the obvious one resulting from restriction of the $\mathfrak{l}_r$-action on $\mathfrak{n}_r^+$ to $\mathfrak{k}_0$. The Lie algebra $E^0(\mathfrak{k})$ is the stabilizer algebra $\mathfrak{z}_{\mathfrak{g}}(e^r)$ of the element e^r generating the principal holomorphic nilpotent orbit $\mathcal{O}_r$, and the stabilizer $\mathrm{Z}_{\mathfrak{g}}(e^r)$ of e^r is the corresponding subgroup of the adjoint group G.

For $\varepsilon > 0$, let $\mathcal{O}^+(\varepsilon) = G(2\varepsilon z)$; these orbits, being semisimple and holomorphic, are ordinary Kähler manifolds, the requisite complex structure being induced from the projection to $\mathfrak{p} \cong \mathfrak{p}^+$, cf. (3.7.6). When ε goes to zero, they tend to or, equivalently, *contract* to the stratified Kähler space $\overline{\mathcal{O}_r}$ (the closure of the principal holomorphic nilpotent orbit $\mathcal{O}_r$) in such a way that the complex structure remains fixed; indeed, the projection to $\mathfrak{p}^+$ is a diffeomorphism from $\mathcal{O}^+(\varepsilon)$ onto $\mathfrak{p}^+$ for $\varepsilon > 0$ and a homeomorphism from $\overline{\mathcal{O}_r}$ onto $\mathfrak{p}^+$ in the limit case $\varepsilon = 0$. Writing $(C^\infty, \{\cdot,\cdot\})_\varepsilon$ $(\varepsilon > 0)$ for the corresponding smooth symplectic Poisson structure on $\mathfrak{p}^+$ and $(C^\infty, \{\cdot,\cdot\})_0$ for the stratified symplectic Poisson structure on $\mathfrak{p}^+$ resulting from the projection from $\overline{\mathcal{O}_r}$ onto $\mathfrak{p}^+$, we obtain a 1-parameter family $(C^\infty, \{\cdot,\cdot\})_\varepsilon$ $(\varepsilon \geq 0)$ of Poisson structures on $\mathfrak{p}^+$, each of which is compatible with the complex structure on $\mathfrak{p}^+$ and thus a complex analytic stratified Kähler structure, an ordinary smooth Kähler structure for $\varepsilon > 0$. *This family is therefore a resolution of singularities* for the stratified Kähler structure on $\overline{\mathcal{O}_r}$, with deformation parameter ε. See (3.2.2) for a very special case. We have defined $\mathcal{O}^+(\varepsilon)$ to be $G(2\varepsilon z)$ (rather than $G(\varepsilon z)$) since, as explained in (5.6), symplectic reduction then identifies $\mathcal{O}^+(1)$ with the homogeneous space G/K in a natural fashion; see also (3.2.2) and (3.4) above.

By Theorem 5.3.3, in each of the standard cases ($\mathfrak{g} = \mathfrak{sp}(\ell, \mathbb{R}), \mathfrak{su}(p,q), \mathfrak{so}^*(2n)$), the closure $\overline{\mathcal{O}_r}$ of the principal holomorphic nilpotent orbit arises via Kähler reduction relative to the corresponding dual group H (which equals, respectively, $\mathrm{O}(\ell, \mathbb{R})$, $\mathrm{U}(p+q, \mathbb{R})$, $\mathrm{Sp}(n)$). The semisimple holomorphic orbits $\mathcal{O}^+(\varepsilon)$ arise as well by reduction relative to the dual group, but with reference to the corresponding element $\varepsilon J_V \in \mathfrak{h}$, as explained in (5.6) above and, in this fashion, inherit Kähler structures which coincide with the Kähler structures whose underlying complex analytic structure is induced from the projection to the constituent $\mathfrak{p}$ in the Cartan decomposition of $\mathfrak{g}$. When the deformation parameter ε tends to zero, $\varepsilon J_V \in \mathfrak{h}$ tends to the origin; this shows once again how the Kähler manifolds $\mathcal{O}^+(\varepsilon)$ contract to the stratified Kähler space $\overline{\mathcal{O}_r}$ in such a way that the complex structure remains fixed.

Remark. For $\varepsilon > 0$, the smooth Kähler structure on the orbit $\mathcal{O}^+(\varepsilon)$ (arising from the projection to $\mathfrak{p}$) is *not* the corresponding hermitian symmetric space structure; in particular, only K acts by isometries, the curvature tensor is not parallel, and the stereographic projection to the corresponding symmetric domain (cf. 3.2.2) is only a symplectic change of coordinates, not a holomorphic one. Moreover, we conjecture that the reasoning in (3.2.2) extends to the general case to the effect that the sectional curvature of the hermitian connection on $\mathcal{O}^+(\varepsilon)$ is negative and that, in the limit, the hermitian connection of each stratum of $\overline{\mathcal{O}_r}$ is flat, i. e. has zero curvature.

10. Projectivization and exotic projective varieties

Let E be a unitary representation of a compact Lie group H, E being endowed with the standard symplectic form. The H-action on E is well known to be hamiltonian and to descend to a hamiltonian H-action on the complex projective space $\mathrm{P}(E)$ arising from E by projectivization, $\mathrm{P}(E)$ being endowed with the appropriate symplectic form. One way to see this is to observe that the central copy of the circle group S^1 in the unitary group $\mathrm{U}(E)$ of E acts on E in a hamiltonian fashion, with a momentum mapping which in physics arises as the *harmonic oscillator* energy hamiltonian. Symplectic reduction with respect to a constant positive value then yields $\mathrm{P}(E)$ endowed with the appropriate symplectic form (a multiple of the negative of the imaginary part of the Fubini-Study metric), and the H-momentum mapping on E passes to a H-momentum mapping on $\mathrm{P}(E)$.

We now apply this observation to the circumstances of Theorem 5.3.3 and maintain the notation established there. Thus, $\mathfrak{g}$ is one of the classical Lie algebras $\mathfrak{sp}(\ell,\mathbb{R})$, $\mathfrak{su}(p,q)$ $(p \geq q)$, $\mathfrak{so}^*(2n)$, and r denotes its real rank.

The hamiltonian action of $H = \mathrm{U}(V^s,(\cdot,\cdot))$ on $W_J = \mathrm{Hom}_{\mathbb{K}}(V^s,V)$ descends to a hamiltonian action on $\mathrm{P}(W_J)$, and the latter carries a smooth (positive) Kähler structure; we remind the reader that the requisite momentum mapping on W_J is that denoted by μ_{H}. in Section 5 above. For $1 \leq s \leq r$, let Q_s be the resulting H-reduced space; in view of Proposition 4.2, since $\mathrm{P}(W_J)$ is a Kähler manifold, Q_s inherits a (positive) normal Kähler structure whose (stratified symplectic) Poisson structure is determined by a choice of positive (energy) value η for the S^1-momentum mapping. The next result refers to the standard cases (cf. (3.5) above).

THEOREM 10.1. *Each Q_s inherits a normal Kähler structure in such a way that*
(i) *the injections $Q_1 \subseteq Q_2 \subseteq \cdots \subseteq Q_r$ are morphisms of (complex analytic) stratified Kähler spaces; that*
(ii) *as a complex analytic space, Q_r is the complex projective space $\mathrm{P}(\mathfrak{p}^+)$; and that*
(iii), *for $s < r$, as a complex analytic space, Q_s is a projective determinantal variety in Q_r.*

PROOF. This is an immediate consequence of Theorem 5.3.3, combined with Proposition 4.2. □

Even though, as a complex analytic space, Q_r is ordinary complex projective space (of complex dimension $\dim(\mathfrak{p}^+) - 1$) and hence non-singular, as a *stratified symplectic space*, Q_r has *singularities*, the *singular locus in the sense of stratified symplectic spaces* being the subspace Q_{r-1}. More precisely, the reduced Poisson bracket is symplectic on the top stratum of Q_r but has smaller rank at every point of Q_{r-1}. Hence the reduced Poisson structure on Q_r does *not* amount to the standard symplectic Poisson structure on complex projective space. We refer to such a projective space which is endowed with a normal Kähler structure whose underlying stratified symplectic structure is *not* standard as an *exotic* projective space, and we call the closure of a stratum an *exotic projective Kähler variety*. The above construction yields the exotic projective spaces $Q_r = \mathrm{P}(\mathfrak{p}^+) = \mathbb{C}\mathrm{P}^d$ according to the three standard cases $(\mathfrak{g},r,d) = (\mathfrak{sp}(\ell,\mathbb{R}),\ell,\frac{1}{2}\ell(\ell+1)-1)$, $(\mathfrak{g},r,d) = (\mathfrak{su}(p,q),q,pq-1)$ $(p \geq q)$, $(\mathfrak{g},r,d) = (\mathfrak{so}^*(2n),[\frac{n}{2}],\frac{1}{2}n(n-1)-1)$, and in each case, the closures of the strata constitute a sequence $Q_1 \subseteq Q_2 \subseteq \cdots \subseteq Q_r$ of exotic projective normal Kähler varieties which, as projective varieties, are determinantal varieties and thus quite explicitly given.

THEOREM 10.2. *As a stratified symplectic subspace of Q_r, each Q_s is a determinantal subset of Q_r in the sense that it may be described by finitely many real homogeneous determinantal equations.*

PROOF. This is a consequence of Theorem 5.4.1. We leave the details to the reader. □

REMARK 10.3. For the case where $\mathfrak{g} = \mathfrak{sp}(\ell, \mathbb{R})$, as a stratified symplectic space, for $1 \leq s \leq \ell = r$, Q_s is the reduced phase space of a system of ℓ harmonic oscillators in $\mathbb{R}^s$ with constant energy and total angular momentum zero, and the injection of Q_{s-1} into Q_s amounts to such a system in $\mathbb{R}^{s-1}$ being viewed as a system in $\mathbb{R}^s$ via the standard inclusion of $\mathbb{R}^{s-1}$ into $\mathbb{R}^s$. Furthermore, for $s > \ell$, the reduced phase space of a system of ℓ harmonic oscillators in $\mathbb{R}^s$ with constant energy and total angular momentum zero is, as a stratified symplectic space, diffeomorphic to the reduced phase space Q_ℓ of a system of ℓ harmonic oscillators in $\mathbb{R}^\ell$ with constant energy and total angular momentum zero, the diffeomorphism being induced by the standard inclusion of $\mathbb{R}^\ell$ into $\mathbb{R}^s$. This diffeomorphism is an isomorphism of (complex analytic) stratified Kähler spaces.

Under the circumstances of Theorem 10.1, the space Q_s arises as well from the closure $\overline{\mathcal{O}_s}$ of the corresponding holomorphic nilpotent orbit $\mathcal{O}_s$ by *stratified symplectic reduction* in the sense explained in Section 4 above. This observation carries over to *all* Lie algebras of hermitian type, in the following fashion: Let $(\mathfrak{g}, z)$ be a simple Lie algebra of hermitian type, with Cartan decomposition $\mathfrak{g} = \mathfrak{k} \oplus \mathfrak{p}$. As before, let G be the adjoint group of $\mathfrak{g}$ and K the maximal compact subgroup thereof with $\mathfrak{k} = \mathrm{Lie}(K)$. For $1 \leq s \leq r$, consider $\overline{\mathcal{O}_s}$, endowed with the complex analytic stratified Kähler structure given in Subsection 3.7. The (co)adjoint action of G on $\mathfrak{g} \cong \mathfrak{g}^*$ induces a G-action on $\overline{\mathcal{O}_s}$ preserving the stratified symplectic structure, and the injection $\overline{\mathcal{O}_s} \subseteq \mathfrak{g}^*$ is a stratified symplectic space momentum mapping for this action. The restriction of this action to K preserves the complex analytic stratified Kähler structure, and the composite of the inclusion with the projection $\mathfrak{g}^* \longrightarrow \mathfrak{k}^*$ is a stratified symplectic space momentum mapping $\overline{\mathcal{O}_s} \longrightarrow \mathfrak{k}^*$ for this action. In particular, the central copy of the circle group S^1 in $K \subseteq G$ acts in this way on $\overline{\mathcal{O}_s}$ in a hamiltonian fashion; since the 1-parameter subgroup of $G = \mathrm{Ad}(\mathfrak{g})$ generated by z is this central copy of S^1 (it consists of the points $\exp(tz) \in G$ with $0 \leq t < 2\pi$ where $\exp\colon \mathfrak{g} \to G$ is the exponential map for the adjoint group G), appropriately scaled, the momentum mapping $\overline{\mathcal{O}_s} \longrightarrow \mathbb{R} \cong (\mathrm{Lie}(S^1))^*$ for this S^1-action is simply given by the composite of the inclusion of $\overline{\mathcal{O}_s}$ into $\mathfrak{g}^*$, followed by z, viewed as a linear map from $\mathfrak{g}^*$ to the reals. For example, for $\mathfrak{g} = \mathfrak{sl}(2, \mathbb{R})$, $\exp(tz) = \rho(\mathrm{e}^{tz})$ where $\mathrm{e}^{tz} = \begin{bmatrix} \cos\frac{t}{2} & -\sin\frac{t}{2} \\ \sin\frac{t}{2} & \cos\frac{t}{2} \end{bmatrix} \in \mathrm{SL}(2, \mathbb{R})$ and where $\rho\colon \mathrm{SL}(2, \mathbb{R}) \longrightarrow \mathrm{PSL}(2, \mathbb{R})$ is the canonical projection. Thus, in general, for each $1 \leq s \leq r$, stratified symplectic reduction with respect to the appropriate positive (energy) value η then yields a complex analytic stratified Kähler space Q_s, with its stratified symplectic Poisson structure determined by the choice of η; complex analytically, the space $\overline{\mathcal{O}_s}$ is the (affine complex) cone on Q_s. In the standard cases, the resulting spaces Q_s coincide with those described in Theorem 10.1 (and denoted by Q_s as well).

For the remaining classical case $\mathfrak{g} = \mathfrak{so}(2, q)$ $(q \geq 3)$, this construction leads to an exceedingly attractive result: Recall from Theorem 3.6.2 that $\mathfrak{so}(2, q)$ has two non-trivial holomorphic nilpotent orbits $\mathcal{O}_1$ and $\mathcal{O}_2$ whose closures constitute a sequence $0 \subseteq \overline{\mathcal{O}_1} \subseteq \overline{\mathcal{O}_2} \cong \mathbb{C}^q$ of normal Kähler spaces; here $\overline{\mathcal{O}_1}$ is the complex quadric $w_1^2 + \cdots + w_q^2 = 0$ in $\overline{\mathcal{O}_2} = \mathbb{C}^q$. Now $G = \mathrm{SO}(2, q)^0$, $K = \mathrm{SO}(2, \mathbb{R}) \times \mathrm{SO}(q, \mathbb{R})$, and $\mathrm{SO}(2, \mathbb{R})$ is the requisite central copy of the circle group in K. By construction, the S^1-reduced space Q_1 sits inside the S^1-reduced space $Q_2 = \mathbb{C}\mathrm{P}^{q-1}$ as the projective quadric in $\mathbb{C}\mathrm{P}^{q-1}$ given by the homogeneous equation $w_1^2 + \cdots + w_q^2 = 0$. This quadric, being a complex analytic stratified Kähler space with a single stratum, is necessarily a smooth Kähler manifold. Summing up, we obtain the following.

THEOREM 10.4. *For $m \geq 2$, $\mathbb{C}\mathrm{P}^m$ carries an exotic normal Kähler structure whose singular locus (in the sense of stratified Kähler spaces) is a quadric Q in $\mathbb{C}\mathrm{P}^m$ in such a way that the exotic Kähler structure on $\mathbb{C}\mathrm{P}^m$ restricts to an ordinary Kähler structure on Q.* □

The exceptional cases lead again to *new* geometric insight: Recall from Section 8 that $\mathfrak{g} = \mathfrak{e}_{7(-25)}$ has three non-trivial holomorphic nilpotent orbits $\mathcal{O}_1$, $\mathcal{O}_2$, and $\mathcal{O}_3$ whose closures constitute a sequence $0 \subseteq \overline{\mathcal{O}_1} \subseteq \overline{\mathcal{O}_2} \subseteq \overline{\mathcal{O}_3}$ of complex analytic stratified Kähler spaces; furthermore, under the projection from $\mathfrak{g}$ to $\mathfrak{p}$, followed by the complex linear isomorphism from $\mathfrak{p}$ onto $\mathfrak{p}^+$, as a stratified space, the closure $\overline{\mathcal{O}_3}$ of the principal holomorphic nilpotent orbit is identified with $\mathcal{H}_3(\mathbb{O}_\mathbb{C})$. Now $G = \mathrm{E}_{7(-25)}$, K is locally isomorphic to $\mathrm{E}_{6(-78)} \times \mathrm{SO}(2, \mathbb{R})$, and the constituent $\mathrm{SO}(2, \mathbb{R})$ is the requisite central copy of the circle group in K (here "requisite" is intended to refer to the fact that, in order for K to be the stabilizer of a point of a symmetric domain, it must have a 1-dimensional center). By construction, the S^1-reduced space Q_2 (here and below "reduced" is understood in the sense of stratified symplectic reduction) sits inside the S^1-reduced space $Q_3 = \mathrm{P}(\mathcal{H}_3(\mathbb{O}^\mathbb{C})) \cong \mathbb{C}\mathrm{P}^{26}$ as the projective complex generic norm (cubic) hypersurface; it is the projectivization of the affine Freudenthal cubic $\mathcal{F}$ mentioned in Section 8 above. Furthermore, the S^1-reduced space Q_1 sits inside Q_2 as its singular locus in the sense of stratified Kähler spaces. Thus the projective cubic Q_2 inherits a complex analytic stratified Kähler structure with two strata. We note that the generic norm hypersurface has been shown by Jacobson to be rational [**48**]. Summing up, we obtain the following.

THEOREM 10.5. *The complex projective space of dimension 26 carries an exotic Kähler structure whose singular locus (in the sense of stratified Kähler spaces) is the projective generic norm (cubic) hypersurface, and the latter inherits a complex analytic stratified Kähler structure with two strata, having as singular stratum Q_1 (in the sense of stratified Kähler spaces) the projectivization $\mathrm{P}(\widetilde{\mathcal{O}_1}) \subseteq \mathrm{P}(\mathfrak{p}^+) \cong \mathbb{C}\mathrm{P}^{26}$ of the lowest non-zero stratum $\widetilde{\mathcal{O}_1} \subseteq \mathcal{H}_3(\mathbb{O}_\mathbb{C})$ (i. e. Jordan rank 1 stratum or, equivalently, lowest non-zero $K^\mathbb{C}$-stratum). The latter amounts to the compact hermitian symmetric space $\mathrm{Ad}(\mathfrak{e}_{6(-78)})/(\mathrm{SO}(10, \mathbb{R}) \cdot \mathrm{SO}(2, \mathbb{R}))$ of complex dimension 16, and the exotic Kähler structure on $\mathbb{C}\mathrm{P}^{26}$ restricts to an ordinary Kähler structure on Q_1.* □

Likewise $\mathfrak{e}_{6(-14)}$ has two non-trivial holomorphic nilpotent orbits $\mathcal{O}_1$ and $\mathcal{O}_2$ whose closures constitute a sequence $0 \subseteq \overline{\mathcal{O}_1} \subseteq \overline{\mathcal{O}_2} \cong \mathfrak{p}^+$ of complex analytic stratified Kähler spaces, and the pre-homogeneous space structure on $\mathfrak{p}^+ \cong \mathbb{C}^{16}$ restricts to the positive half-spin representation $\mathcal{S}_+^\mathbb{C}$ of $\mathrm{Spin}(10, \mathbb{C})$ (cf. Section 8 above for the notation).

THEOREM 10.6. *The complex projective space of dimension* 15 *carries an exotic Kähler structure whose singular locus* Q *(in the sense of stratified Kähler spaces) is the projectivization* $\mathrm{P}(\widetilde{\mathcal{O}_1}) \subseteq \mathrm{P}(\mathfrak{p}^+) \cong \mathbb{C}\mathrm{P}^{15}$ *of the space* $\widetilde{\mathcal{O}_1}$ *of pure spinors in* $\mathfrak{p}^+ \cong \mathcal{S}_+^{\mathbb{C}}$, *and* Q *amounts to the compact hermitian symmetric space* $\mathrm{SO}(10,\mathbb{R})/\mathrm{U}(5)$ *of complex dimension* 10. *The exotic Kähler structure on* $\mathbb{C}\mathrm{P}^{15}$ *restricts to an ordinary Kähler structure on* Q. □

Indeed, cf. e. g. [**59**] (IV.9.7 p. 337), it is well known that the projective space associated with the pure spinors for the (16-dimensional) positive half-spin representation $\mathcal{S}_+^{\mathbb{C}}$ of $\mathrm{Spin}(10,\mathbb{C})$ is the compact hermitian symmetric space $\mathrm{SO}(10,\mathbb{R})/\mathrm{U}(5)$. For intelligibility, we recall that, by Theorem 8.5.3, under the homeomorphism from $\overline{\mathcal{O}_2}$ to $\mathfrak{p}^+$, the orbit $\mathcal{O}_1$ is identified with the space $\widetilde{\mathcal{O}_1}$ of pure spinors in $\mathfrak{p}^+ \cong \mathcal{S}_+^{\mathbb{C}}$.
10.7. PROJECTIVE NORMALITY. A projective variety $X \subseteq \mathbb{C}\mathrm{P}^d$ is said to be *projectively normal* (for the given embedding) provided its homogeneous coordinate ring is integrally closed. This is equivalent to requiring that the cone on X (in $\mathbb{C}^{d+1}$) be normal as an affine variety. In the classical cases since, for $1 \leq s \leq r$, complex analytically, the space $\overline{\mathcal{O}_s}$ is a normal affine variety (Theorems 3.5.5, 3.6.3, 5.3.3), each Q_s is projectively normal. Furthermore, cf. Theorem 8.4.2, the affine Freudenthal cubic is normal whence the projective Freudenthal cubic is projectively normal. We now settle the remaining cases, the lowest non-trivial stratum in each of the two exceptional cases. We are indebted to L. Manivel for having communicated to us the following.

PROPOSITION 10.7.1. *The closed orbit* X *in* $\mathrm{P}(V)$ *of an irreducible complex representation* V *of a complex simple Lie group is projectively normal.*

PROOF. Indeed, in view of the Borel-Weil theorem, if v is the highest weight of the irreducible representation $V^* = \Gamma(X, \mathcal{O}_X(1))$, for $n \geq 1$, the highest weight of the irreducible representation $V^* = \Gamma(X, \mathcal{O}_X(n))$ equals nv. Consequently for $n \geq 1$ the natural map $\Gamma(\mathrm{P}(V), \mathcal{O}_{\mathrm{P}(V)}(n)) \to \Gamma(X, \mathcal{O}_X(n))$ is surjective whence, cf. Ex. II.5.14 on p. 126 of [**21**], the projective variety X is projectively normal. □

Given a simple Lie algebra of hermitian type $(\mathfrak{g}, z)$, we may apply this Proposition to the irreducible representation of $K^{\mathbb{C}}$ on $\mathfrak{p} \cong \mathfrak{p}^+$, the closed orbit X being the orbit denoted earlier by Q_1. Thus we conclude that Q_1 is projectively normal (which we already know in the classical cases). Consequently, for $\mathfrak{e}_{6(-14)}$, the projectivization $Q_1 = \mathrm{P}(\widetilde{\mathcal{O}_1}) \subseteq \mathrm{P}(\mathfrak{p}^+) \cong \mathbb{C}\mathrm{P}^{15}$ of the space $\widetilde{\mathcal{O}_1}$ of pure spinors in $\mathcal{S}_+^{\mathbb{C}}$ is projectively normal (for the embedding into $\mathrm{P}(\mathfrak{p}^+)$) whence, complex analytically, $\overline{\mathcal{O}_1}$ is a normal affine variety. Likewise, for $\mathfrak{e}_{7(-25)}$, the projectivization Q_1 of $\overline{\mathcal{O}_1}$ is projectively normal (for the embedding into $Q_3 = \mathrm{P}(\mathcal{H}_3(\mathbb{O}^{\mathbb{C}})) \cong \mathbb{C}\mathrm{P}^{26}$) whence $\overline{\mathcal{O}_1}$ is normal as an affine complex variety.

REMARK 10.8. A Kodaira embedding of a smooth projective variety N carrying a Kähler structure will in general not be compatible with the symplectic structures, where projective space is endowed with the Fubini-Studi metric. Furthermore, given an arbitrary smooth symplectic manifold X and a smooth symplectic submanifold Y, the symplectic Poisson algebra on X will *not* restrict to the symplectic Poisson algebra on Y; indeed this observation prompted DIRAC to introduce what is nowadays called the *Dirac bracket*. The above results indicate that, perhaps, the right question to ask is this: Given N and an embedding thereof into some complex projective space is there an exotic Kähler structure on that projective space which restricts to the Kähler structure on N? More generally, we could ask this question for an arbitray compact complex analytic stratified Kähler space N, and the above examples show that this situation indeed occurs. In particular: A moduli space of the kind explained in Example 4.3 above is known to be complex analytically a normal projective variety. Thus we are led to the following question: Given such a space N, *is there an embedding of N into some complex projective space endowed with an exotic Kähler structure such that this exotic Kähler structure restricts to the normal Kähler structure on N* explained in Example 4.3?

11. Comparison with other notions of Kähler space with singularities

Let X be a reduced complex analytic space (where "reduced" is understood in the sense of complex analytic spaces, *not* in the sense of symplectic geometry), let N be its singular locus, and consider a hermitian metric g on $X \setminus N$. In [**16**] (p. 346) GRAUERT refers to such a metric as a *Kähler metric* on X provided every point x of X has a neighborhood U and a strongly plurisubharmonic function p defined on U in such a way that, on $U \setminus N$, the metric g coincides with the Levi form of p, that is to say, in suitable coordinates $z_1, \ldots, z_m$,

$$g_{j,\overline{k}} = \frac{\partial^2 p}{\partial z_j \partial \overline{z}_k}.$$

As usual, the function p is then referred to as a *local Kähler potential* for g.

On the other hand, given a stratified Kähler space of complex dimension n, on its top stratum, in local coordinates $z_1, \ldots, z_m$, the Poisson structure and Kähler metric g satisfy the identity

$$\sum_{\ell} \{\overline{z}_k, z_\ell\} g_{\ell,\overline{m}} = 2i\delta_{k,m}, \quad \text{for every} \quad 1 \leq k, m \leq n,$$

that is, the functions $\frac{1}{2i}\{\overline{z}_k, z_\ell\}$ yield the coefficients of the inverse matrix function g^{-1} (in these coordinates). Since the Poisson structure is defined on the whole space, even though the Kähler metric is *not*, we can somewhat loosely interpret the above formula by saying that the "inverse" of the Kähler metric *is* defined, in fact, even in such a way that the restriction of this "inverse" to any lower stratum is there invertible so that the inverse of the "inverse", in turn, defines a Kähler metric on the stratum as well. The Poisson structure then encapsulates the mutual positions of these Kähler structures.

Grauert's notion has been refined in [**88**], cf. also [**22**], to that of a *stratified Kählerian space*; such a structure encapsulates the mutual positions of Kähler structures on pieces of a decomposition, but in a way different from our notion of complex analytic stratified Kähler space. To reproduce stratified Kählerian spaces briefly, let X be a reduced complex analytic space, with a complex analytic stratification by complex analytic manifolds (in general genuinely finer than the ordinary complex analytic stratification). Given such a space, a *Kählerian* structure on X is a collection of "strictly plurisubharmonic functions" $\{\phi_\alpha\}$ which are defined and continuous on the constituents U_α of an open covering $\{U_\alpha\}$ of X and smooth on the intersection of each stratum with U_α in such a way that the differences $u_{\alpha,\beta}$ are pluriharmonic in the strict sense on the intersections $U_{\alpha,\beta} = U_\alpha \cap U_\beta$, that is, there is $f_{\alpha,\beta} \in \mathcal{O}(U_{\alpha,\beta})$ such that $u_{\alpha,\beta} = \operatorname{Re}(f_{\alpha,\beta})$. There are various ways to make precise the idea of strictly plurisubharmonic function on X: the strongest one is the notion of "strictly plurisubharmonic in the sense of perturbations" (which we do not reproduce here) and a somewhat weaker on is that of "stratum-wise strictly plurisubharmonic" which means that the restriction to (the appropriate open subset of) any stratum of X is smooth in such a way that its Levi-form is positive definite (i. e. a hermitian structure); see [**22**] for details.

To explain the relationship between stratified Kähler spaces and stratified Kählerian spaces, let $(X, C^\infty(X), \{\cdot,\cdot\})$ be a stratified symplectic space. Write $A = C^\infty(X)$, let $(A, L) = (C^\infty(X), \operatorname{Der}(A))$, with its Lie-Rinehart algebra structure, and consider the differential graded algebra $(\operatorname{Alt}_A(L, A), d)$ determined by it

[**29**]; when X is a smooth manifold and $C^\infty(X)$ the algebra of ordinary smooth functions, $(\mathrm{Alt}_A(L,A),d)$ is the differential graded algebra of ordinary differential forms for X. For a general stratified symplectic space $(X, C^\infty(X), \{\cdot,\cdot\})$, the Poisson structure induces a morphism $\pi^\sharp_{\{\cdot,\cdot\}}\colon (A, \Omega^1(X)_{\{\cdot,\cdot\}}) \to (A, L)$ of Lie-Rinehart algebras (cf. (1.1) above) and hence a morphism

$$(\mathrm{Alt}_A(L,A),d) \longrightarrow (\mathrm{Alt}_A(\Omega^1(X)_{\{\cdot,\cdot\}},A), d_{\{\cdot,\cdot\}})$$

of differential graded algebras [**29**]; the Poisson 2-form $\pi_{\{\cdot,\cdot\}}$ is then an element of $\mathrm{Alt}^2_A(\Omega^1(X)_{\{\cdot,\cdot\}},A)$, cf. Section 1 above.

On the other hand, consider a stratified Kählerien space Y, with its ordinary complex analytic stratification. The latter determines a smooth structure $C^\infty_{\mathrm{min}}(Y)$ which we refer to as the *minimal one*; this structure is, roughly speaking, locally determined by the embedding of the corresponding complex analytic germ into a complex domain [**88**]; the requisite smooth structure on Y is then obtained (locally) by restricting the ordinary smooth structure on the domain. (Varouchas explains this in terms of sheaves.) This smooth structure determines a (complex) de Rham algebra $\Omega_Y = \Lambda\Omega^1_Y$, Ω^1_Y being defined by an exact sequence of the kind

$$I/I^2 \longrightarrow C^\infty_{\mathrm{min}}(Y,\mathbb{C}) \otimes_{C^\infty(X,\mathbb{C})} \Omega^1_X \longrightarrow \Omega^1_Y \longrightarrow 0$$

where $Y \subseteq X$ is an embedding into a complex domain X and where I is the ideal (sheaf) of (germs of) functions which vanish on Y. The complex analytic structure of Y entails a bigrading $\Omega^{*,*}_Y$, just as for ordinary smooth complex manifolds, and the stratified Kählerien structure determines a 2-form $\omega \in \Omega^{1,1}_Y$ which, when Y is an ordinary smooth Kähler manifold, boils down to its standard symplectic form.

Return to the circumstances of Proposition 4.2. The reduced space N^{red} coming into play there inherits, on the one hand, a smooth structure $C^\infty(N^{\mathrm{red}},\mathbb{R})$ which underlies its stratified symplectic Poisson algebra and, furthermore, a complex analytic stratified Kähler space structure. On the other hand, by (3.4) of [**23**], N^{red} also inherits a stratified Kählerien structure. The corresponding minimal smooth structure $C^\infty_{\mathrm{min}}(N^{\mathrm{red}})$ (explained above) will in general be a proper subalgebra of $C^\infty(N^{\mathrm{red}},\mathbb{R})$, and the space $\Omega^1_{N^{\mathrm{red}}}$ of 1-forms with reference to the complexified algebra $C^\infty_{\mathrm{min}}(N^{\mathrm{red}},\mathbb{C})$ will decompose into a direct sum of $\Omega^{0,1}_{N^{\mathrm{red}}}$ and $\Omega^{1,0}_{N^{\mathrm{red}}}$. Notice that, cf. Remark 3.7.3, a general complex analytic stratified Kähler structure does *not* give rise to such a decomposition; indeed, cf. (3.7.4), the sum of P and $\overline{P}$ does not exhaust the space of 1-forms. To explore the mutual relationship between the complex analytic stratified Kähler and stratified Kählerien structures (if any), write $A = C^\infty(N^{\mathrm{red}},\mathbb{C})$ and $L = \mathrm{Der}(A)$), and consider the differential graded algebra $(\mathrm{Alt}_A(L,A),d)$ determined by it, that is, the Rinehart complex for the Lie-Rinehart algebra $(C^\infty(N^{\mathrm{red}},\mathbb{C}), \mathrm{Der}(A))$. With reference to the ordinary complex analytic stratification, the stratified Kählerien structure determines a 2-form $\omega \in \Omega^{1,1}_{N^{\mathrm{red}}} \subseteq \Omega^2_{N^{\mathrm{red}}}$ whereas the complex analytic stratified Kähler structure determines the 2-form $\pi_{\{\cdot,\cdot\}} \in \mathrm{Alt}^2_A(\Omega^1(N^{\mathrm{red}})_{\{\cdot,\cdot\}},A)$. In the case of an ordinary complex Kähler manifold, we can identify ω with $\pi_{\{\cdot,\cdot\}}$ but for a general stratified Kählerien structure (i. e. having underlying stratification finer than the ordinary complex analytic one), there is no obvious replacement for the constituent $\Omega^{1,1}_{N^{\mathrm{red}}}$ and hence no obvious candidate for a 2-form of the kind ω whereas given a complex analytic stratified Kähler structure, the 2-form $\pi_{\{\cdot,\cdot\}}$ is always defined. Furthermore, when there is a 2-form of the kind ω with reference to the minimal structure $C^\infty_{\mathrm{min}}(N^{\mathrm{red}})$, its image in $\mathrm{Alt}^2_A(\Omega^1(N^{\mathrm{red}})_{\{\cdot,\cdot\}}, A)$ will not in general coincide with the 2-form $\pi_{\{\cdot,\cdot\}}$. To justify this claim, return to the circumstances of Theorem 3.5.4. In view of Theorem 3.5.5, complex analytically, $\overline{\mathcal{O}_r}$ amounts to $\mathfrak{p}^+$ and, for $s < r$, the complex analytic structure on $\overline{\mathcal{O}_s}$ arises from the embedding of $\overline{\mathcal{O}_s}$ into $\mathfrak{p}^+$. Moreover, the minimal smooth structure $C^\infty_{\mathrm{min}}(\overline{\mathcal{O}_s})$ (i. e. algebra of continuous functions) on $\overline{\mathcal{O}_s}$ is determined by the embedding of $\overline{\mathcal{O}_s}$ into $\mathfrak{p}^+$, too; this is the structure coming into play in [**88**]. On the other hand, the stratified symplectic structure of $\overline{\mathcal{O}_s}$ results from its embedding into $\mathfrak{g} \cong \mathfrak{g}^*$; a choice of basis of $\mathfrak{g}$ determines a system of coordinate functions on $\overline{\mathcal{O}_s}$, and the last statement of Theorem 5.4.1 says that the stratified symplectic Poisson algebra on $\overline{\mathcal{O}_s}$ cannot be generated by fewer functions. Furthermore, $C^\infty_{\mathrm{min}}(\overline{\mathcal{O}_s})$ being viewed as a subalgebra of $C^\infty(\overline{\mathcal{O}_s}, \mathbb{C})$, Poisson brackets of elements of $C^\infty_{\mathrm{min}}(\overline{\mathcal{O}_s})$ do not necessarily lie in $C^\infty_{\mathrm{min}}(\overline{\mathcal{O}_s})$. Summing up, we see that the stratified symplectic structure of $\overline{\mathcal{O}_s}$ results from its embedding into $\mathfrak{g}$ whereas its complex analytic structure comes from the composite of this embedding with the projection from $\mathfrak{g}$ to $\mathfrak{p}$. Even though the projection from $\mathfrak{g}$ to $\mathfrak{p}$ is *not* injective, the restriction thereof to $\overline{\mathcal{O}_s}$ is still injective and turns $\overline{\mathcal{O}_s}$ into a complex analytic space. Thus, the real and complex structures of $\overline{\mathcal{O}_s}$ are obtained from the embeddings thereof into $\mathfrak{g}$ and $\mathfrak{p}$, respectively, and the relationship between the two structures comes from the projection from $\mathfrak{g}$ to $\mathfrak{p}$. Perhaps this kind of situation should be taken as a local model for a complex analytic stratified Kähler space in general.

This discussion raises the following questions:

1) Given a complex analytic stratified Kähler structure on a stratified symplectic space X, do there always exist local Kähler potentials which determine a stratified Kählerian structure?

2) Consider a complex analytic space X, endowed with a general stratified Kählerien structure where "general" means that the underlying stratification may be finer than the ordinary complex analytic one, and let $C^\infty(X)$ be the smooth structure whose elements are continuous functions on X which, restricted to each stratum, are smooth. Given two functions $f, h \in C^\infty(X)$, taking their Poisson bracket $\{f, h\}$ stratum-wise, we obtain a function on X which is necessarily smooth on every stratum. The question is whether $\{f, h\}$ yields a *continuous* function on X; if so, $C^\infty(X)$ would be closed under the Poisson bracket and, in view of Theorem 2.5, the latter would turn X into a complex analytic stratified Kähler space.

The answers to both questions are "yes" for a space X which arises by symplectic reduction from a (positive) Kähler manifold with respect to a compact group. Thus to find a potential counterexample to (1), we would have to look for a complex analytic stratified Kähler space which does not arise in this way from an ordinary Kähler manifold; for $q \geq 5$ *the exotic complex projective space described in* Theorem 10.4 *above having as its singular locus a quadric seems indeed to be such an example, and we do not know whether there exist local Kähler potentials in this case which determine a stratified Kählerian structure*. The same kind of question can be asked for the two exceptional cases $\mathfrak{e}_{6(-14)}$ and $\mathfrak{e}_{7(-25)}$. Likewise, for a potential counterexample to (2), we would need a stratified Kählerien space which does not arise from an ordinary Kähler manifold by reduction.

References

1. J. M. Arms, R. Cushman, and M. J. Gotay, *A universal reduction procedure for Hamiltonian group actions*, in: The geometry of Hamiltonian systems, T. Ratiu, ed., MSRI Publ. **20** (1991), Springer Verlag, Berlin · Heidelberg · New York · Tokyo, 33–51.
2. J. M. Arms, J. E. Marsden, and V. Moncrief, *Symmetry and bifurcation of moment mappings*, Comm. Math. Phys. **78** (1981), 455–478.
3. M. F. Atiyah and R. Bott, *The Yang-Mills equations over Riemann surfaces*, Phil. Trans. R. Soc. London A **308** (1982), 523–615.
4. D. Barbasch and M. R. Sepanski, *Closure ordering and the Kostant-Sekiguchi correspondence*, Proc. Amer. Math. Soc. **126** (1998), 311–317.
5. A. F. Beardon, *The Geometry of Discrete Groups*, Graduate Texts in Mathematics, Springer Verlag, Berlin · Heidelberg · New York · Tokyo, 1983.
6. W. Bruns and U. Vetter, *Determinantal Rings*, Lecture Notes in Mathematics, Vol. 1327, Springer Verlag, Berlin · Heidelberg · New York, 1988.
7. N. Bourgoyne and R. Cushman, *Conjugacy classes in linear groups*, J. of Algebra **44** (1977), 339–362.
8. C. Chevalley, *Invariants of finite groups generated by reflections*, Amer. J. of Math. **77** (1955), 778–782.
9. D. H. Collingwood and W. M. McGovern, *Nilpotent orbits in semisimple Lie algebras*, Mathematics Series, Van Nostrand Reinhold, New York, 1993.
10. R. Cushman, *The momentum mapping of the harmonic oscillator*, Symposia Math. **14** (1975), 323–342.
11. R. Cushman and J. Śniatycki, *Differential structure of orbit spaces*, Canad. J. Math. **53** (2001), 715–735.
12. D. Djoković, *Classification of nilpotent elements in simple exceptional real Lie algebras of inner type and descriptions of their centralizers*, J. of Algebra **112** (1988), 503–524.
13. G. B. Elkington, *Centralizers of unipotent elements in semisimple algebraic groups*, J. of Algebra **23** (1972), 137–163.
14. J. Faraut and A. Koranyi, *Analysis On Symmetric Cones*, Oxford Mathematical Monographs No. 114, Oxford University Press, Oxford U.K., 1994.
15. H. Freudenthal, *Beziehungen der E_7 und E_8 zur Oktavenebene*, Indagationes Math. **16** (1954), 218–230.
16. H. Grauert, *Über Modifikationen und exzeptionelle analytische Mengen*, Math. Ann. **146** (1962), 331–368.
17. H. Grauert and R. Remmert, *Coherent Analytic Sheaves*, Grundlehren der mathematischen Wissenschaften, vol. 265, Springer Verlag, Berlin · Heidelberg · New York · Tokyo, 1984.
18. V. Guillemin and S. Sternberg, *Symplectic techniques in Physics*, Cambridge University Press, London/New York, 1984.
19. V. W. Guillemin and S. Sternberg, *Geometric quantization and multiplicities of group representations*, Invent. Math. **67** (1982), 515–538.
20. K. Guruprasad, J. Huebschmann, L. Jeffrey, and A. Weinstein, *Group systems, groupoids, and moduli spaces of parabolic bundles*, Duke Math. J. **89** (1997), 377–412, `dg-ga/9510006`.
21. R. Hartshorne, *Algebraic Geometry*, Graduate texts in Mathematics No. 52, Springer Verlag, Berlin-Göttingen-Heidelberg, 1977.
22. P. Heinzner, A. Huckleberry, and F. Loose, *Kählerien extensions of the symplectic reduction*, J. für die reine und angewandte Mathematik **455** (1994), 123–140.
23. P. Heinzner and F. Loose, *Reduction of complex Hamiltonian G-spaces*, Geom. and Func. Analysis **4** (1994), 288–297.
24. S. Helgason, *Differential Geometry, Lie Groups, and Symmetric Spaces*, Pure and Applied Mathematics, A series of monographs and textbooks, Academic Press, New York · London · Toronto · Sydney · San Francisco, 1978.
25. J. Hilgert, K.-H. Neeb, and B. Oersted, *The geometry of nilpotent coadjoint orbits of convex type in hermitian Lie algebras*, J. of Lie Theory **4** (1994), 185–235.
26. M. Hochster and J. A. Eagon, *Cohen-Macaulay rings, invariant theory, and the generic perfection of determinantal loci*, Amer. J. of Math. **93** (1971), 1020–1058.
27. R. Howe, *Remarks on classical invariant theory*, Trans. Amer. Math. Soc. **313** (1989), 539–570.
28. R. Howe, *Transcending classical invariant theory*, J. Amer. Math. Soc. **2** (1989), 535–552.

29. J. Huebschmann, *Poisson cohomology and quantization*, J. für die reine und angewandte Mathematik **408** (1990), 57–113.
30. J. Huebschmann, *On the quantization of Poisson algebras*, Symplectic Geometry and Mathematical Physics, Actes du colloque en l'honneur de Jean-Marie Souriau, P. Donato, C. Duval, J. Elhadad, G.M. Tuynman, eds.; Progress in Mathematics, Vol. 99 (1991), Birkhäuser Verlag, Boston · Basel · Berlin, 204–233.
31. J. Huebschmann, *The singularities of Yang-Mills connections for bundles on a surface. I. The local model*, Math. Z. **220** (1995), 595–605, `dg-ga/9411006`.
32. J. Huebschmann, *The singularities of Yang-Mills connections for bundles on a surface. II. The stratification*, Math. Z. **221** (1996), 83–92, `dg-ga/9411007`.
33. J. Huebschmann, *Poisson geometry of flat connections for* SU(2)*-bundles on surfaces*, Math. Z. **221** (1996), 243–259, `hep-th/9312113`.
34. J. Huebschmann, *Poisson structures on certain moduli spaces for bundles on a surface*, Annales de l'Institut Fourier **45** (1995), 65–91, `dg-ga/9411009`.
35. J. Huebschmann, *Symplectic and Poisson structures of certain moduli spaces*, Duke Math. J. **80** (1995), 737–756, `hep-th/9312112`.
36. J. Huebschmann, *Symplectic and Poisson structures of certain moduli spaces. II. Projective representations of cocompact planar discrete groups*, Duke Math. J. **80** (1995), 757–770, `dg-ga/9412003`.
37. J. Huebschmann, *On the variation of the Poisson structures of certain moduli spaces*, Math. Ann. **319** (2001), 267–310, `dg-ga/9710033`.
38. J. Huebschmann, *Smooth structures on moduli spaces of central Yang-Mills connections for bundles on a surface*, J. of Pure and Applied Algebra **126** (1998), 183–221, `dg-ga/9411008`.
39. J. Huebschmann, *Poisson geometry of certain moduli spaces*, Lectures delivered at the "14th Winter School", Srni, Czeque Republic, January 1994, Rendiconti del Circolo Matematico di Palermo, Serie II **39** (1996), 15–35.
40. J. Huebschmann, *On the Poisson geometry of certain moduli spaces*, in: Proceedings of an international workshop on "Lie theory and its applications in physics", Clausthal, 1995 H. D. Doebner, V. K. Dobrev, J. Hilgert, eds. (1996), World Scientific, Singapore · New Jersey · London · Hong Kong, 89–101.
41. J. Huebschmann, *Singularities and Poisson geometry of certain representation spaces*, in: Quantization of Singular Symplectic Quotients, N. P. Landsman, M. Pflaum, M. Schlichenmaier, eds., Workshop, Oberwolfach, August 1999, Progress in Mathematics, Vol. 198 (2001), Birkhäuser Verlag, Boston · Basel · Berlin, 119–135, `math.dg/0012184`.
42. J. Huebschmann, *Lie-Rinehart algebras, descent, and quantization*, Fields Institute Communications (to appear), `math.SG/0303016`.
43. J. Huebschmann, *Kähler reduction and quantization*, `math.SG/0207166`.
44. J. Huebschmann, *Severi varieties and holomorphic nilpotent orbits*, `math.DG/0206143`.
45. J.-I. Igusa, *A classification of spinors up to dimension twelve*, Amer. J. of Math. **92** (1970), 997–1028.
46. E. Inonu and E.P. Wigner, *On the contraction of groups and their representations*, Proc. Nat. Acad. Sci. USA **39** (1953), 510–524.
47. N. Jacobson, *Exceptional Lie algebras*, Lecture Notes in Pure and Applied Mathematics, Marcel Dekker, New York, 1971.
48. N. Jacobson, *Some projective varieties defined by Jordan algebras*, J. of Algebra **97** (1985), 556–598.
49. G. Kempf and L. Ness, *The length of vectors in representation spaces*, Algebraic geometry, Copenhagen, 1978, Lecture Notes in Mathematics **732** (1978), Springer Verlag, Berlin · Heidelberg · New York, 233–244.
50. F. Kirwan, *Cohomology of quotients in symplectic and algebraic geometry*, Princeton University Press, Princeton, New Jersey, 1984.
51. P. Z. Kobak and A. Swann, *Classical nilpotent orbits as hyperkähler quotients*, Int. J. Math. **7** (1996), 193–210.
52. M. Kodaira, L. Nirenberg and D. C. Spencer, *On the existence of deformations of complex analytic structures*, Ann. of Math. **68** (1958), 450–457.
53. B. Kostant, *The principal three-dimensional subgroup and the Betti numbers of a complex simple Lie group*, Amer. J. of Math. **81** (1959), 973–1032.

54. B. Kostant, *Lie group representations on polynomial rings*, Amer. J. of Math. **85** (1963), 327–404.
55. H. Kraft and C. Procesi, *On the geometry of conjugacy classes in classical groups*, Comment. Math. Helv. **57** (1982), 529–602.
56. I. S. Krasil'shchik, V. V. Lychagin, and A. M. Vinogradov, *Geometry of Jet Spaces and Nonlinear Partial Differential Equations*, Advanced Studies in Contemporary Mathematics, vol. 1, Gordon and Breach Science Publishers, New York, London, Paris, Montreux, Tokyo, 1986.
57. P. B. Kronheimer, *Instantons and the geometry of the nilpotent variety*, J. Diff. Geom. **32** (1990), 473–490.
58. S. Kumar, M. S. Narasimhan, and A. Ramanathan, *Infinite Grassmannian and moduli space of G-bundles*, Math. Ann. **300** (1994), 41-75.
59. H. B. Lawson and M. L. Michelsohn, *Spin Geometry*, Princeton Mathematical Series, vol. 34, Princeton University Press, Princeton, N. J., 1989.
60. E. Lerman, R. Montgomery, and R. Sjamaar, *Examples of singular reduction*, Symplectic Geometry, Warwick, 1990, D. A. Salamon, editor, London Math. Soc. Lecture Note Series, vol. 192 (1993), Cambridge University Press, Cambridge, UK, 127–155.
61. O. Loos, *Jordan pairs*, Lecture Notes in Mathematics, Vol. 460, Springer Verlag, Berlin · Heidelberg · New York, 1975.
62. O. Loos, *Bounded symmetric domains and Jordan pairs*, Math. Lectures, University of California, Irvine, California, 1977.
63. J. Marsden and T. Ratiu, *Reduction of Poisson manifolds*, Lett. in Math. Phys. **11** (1985), 161–170.
64. J. Marsden and A. Weinstein, *Reduction of symplectic manifolds with symmetries*, Rep. on Math. Phys. **5** (1974), 121–130.
65. K. McCrimmon, *Jordan algebras and their applications*, Bull. Amer. Math. Soc. **84** (1978), 612–627.
66. V. Mehta and C. Seshadri, *Moduli of vector bundles on curves with parabolic structure*, Math. Ann. **248** (1980), 205–239.
67. K. Mizuno, *The conjugate classes of unipotent elements of the Chevalley groups of type E_7 and E_8*, Tokyo J. Math. **3** (1980), 391–461.
68. I. Muller, H. Rubenthaler, and G. Schiffmann, *Structures des espaces préhomogènes associés à certaines algèbres de Lie graduées*, Math. Ann. **274** (1986), 95–123.
69. M. S. Narasimhan and C. S. Seshadri, *Stable and unitary vector bundles on a compact Riemann surface*, Ann. of Math. **82** (1965), 540–567.
70. M. S. Narasimhan and S. Ramanan, *Moduli of vector bundles on a compact Riemann surface*, Ann. of Math. **89** (1969), 19–51.
71. M. S. Narasimhan and S. Ramanan, *2θ-linear systems on abelian varieties*, Bombay colloquium (1985), 515–427.
72. E. Neher, *On the classification of Lie and Jordan triple systems*, Comm. Algebra **13** (1985), 2615–2667.
73. E. Neher, *Jordan triple systems by the grid approach*, Lecture Notes in Mathematics, Vol. 1280, Springer Verlag, Berlin · Heidelberg · New York, 1987.
74. L. Ness, *A stratification of the null cone via the moment map*, Amer. J. of Math. **106** (1984), 1231–1329.
75. T. Ohta, *The closures of nilpotent orbits in the classical symmetric pairs and their singularities*, Tôhoku Math. J. **43** (1991), 161–211.
76. J.-P. Ortega and T. Ratiu, *Singular reduction of Poisson manifolds*, Letters in Math. Physics **46** (1998), 359–372.
77. H. P. Petersson and M. L. Racine, *Albert Algebras*, in: Jordan Algebras, Proceedings of the Conference held in Oberwolfach, Germany, August 9 – 15, 1992, W. Kaup, K. McCrimmon, H. P. Petersson, editors (1994), Walter de Gruyter, Berlin · New York, 197–207.
78. G. Rinehart, *Differential forms for general commutative algebras*, Trans. Amer. Math. Soc. **108** (1963), 195–222.
79. I. Satake, *Algebraic structures of symmetric domains*, Publications of the Math. Soc. of Japan, vol. 14, Princeton University Press, Princeton, NJ, 1980.
80. M. Sato and T. Kimura, *A classification of irreducible prehomogeneous vector spaces and their relative invariants*, Nagoya Math. J. **65** (1977), 1–155.

81. G. W. Schwarz, *Smooth functions invariant under the action of a compact Lie group*, Topology **14** (1975), 63–68.
82. G. W. Schwarz, *The topology of algebraic quotients*, In: Topological methods in algebraic transformation groups, Progress in Mathematics, Vol. 80 (1989), Birkhäuser Verlag, Boston · Basel · Berlin, 135–152.
83. J. Sekiguchi, *Remarks on real nilpotent orbits of a symmetric pair*, J. Math. Soc. Japan **39** (1987), 127–138.
84. R. Sjamaar, *Holomorphic slices, symplectic reduction, and multiplicities of representations*, Ann. of Math. **141** (1995), 87–129.
85. R. Sjamaar and E. Lerman, *Stratified symplectic spaces and reduction*, Ann. of Math. **134** (1991), 375–422.
86. P. Slodowy, *Simple singularities and simple algebraic groups*, Lecture Notes in Mathematics, Vol. 815, Springer Verlag, Berlin · Heidelberg · New York, 1980.
87. T. Springer and R. Steinberg, *Conjugacy classes*, in: Seminar on Algebraic Groups and Related Finite Groups; Lecture Notes in Mathematics No. 131 (1970), Springer Verlag, Berlin · Heidelberg · New York, 167–266.
88. J. Varouchas, *Kähler spaces and proper open morphisms*, Math. Ann. **283** (1989), 13–52.
89. M. Vergne, *Instantons et correspondance de Kostant-Sekiguchi*, C. R. Acad. Sci. Paris Série I **320** (1995), 901–906.
90. E. B. Vinberg, *On the classification of the nilpotent elements of graded Lie algebras*, Sov. Math. Dokl. **16** (1975), 1517–1520.
91. A. Weinstein, *The local structure of Poisson manifolds*, J. of Diff. Geom. **18** (1983), 523–557.
92. A. Weinstein, *Poisson structures*, in: E. Cartan et les Mathématiciens d'aujourd'hui, Lyon, 25–29 Juin, 1984, Astérisque, **hors-serie** (1985), 421–434.
93. H. Weyl, *The classical groups*, Princeton University Press, Princeton, New Jersey, 1946.
94. H. Whitney, *Complex analytic varieties*, Addison-Wesley Pub. Comp., Reading, Ma, Menlo Park, Ca, London, Don Mills, Ontario, 1972.
95. N. M. J. Woodhouse, *Geometric quantization*, second edition, Clarendon Press, Oxford, 1991.

Editorial Information

To be published in the *Memoirs*, a paper must be correct, new, nontrivial, and significant. Further, it must be well written and of interest to a substantial number of mathematicians. Piecemeal results, such as an inconclusive step toward an unproved major theorem or a minor variation on a known result, are in general not acceptable for publication. Papers appearing in *Memoirs* are generally longer than those appearing in *Transactions*, which shares the same editorial committee.

As of August 1, 2004, the backlog for this journal was approximately 5 volumes. This estimate is the result of dividing the number of manuscripts for this journal in the Providence office that have not yet gone to the printer on the above date by the average number of monographs per volume over the previous twelve months, reduced by the number of volumes published in four months (the time necessary for preparing a volume for the printer). (There are 6 volumes per year, each containing at least 4 numbers.)

A Consent to Publish and Copyright Agreement is required before a paper will be published in the *Memoirs*. After a paper is accepted for publication, the Providence office will send a Consent to Publish and Copyright Agreement to all authors of the paper. By submitting a paper to the *Memoirs*, authors certify that the results have not been submitted to nor are they under consideration for publication by another journal, conference proceedings, or similar publication.

Information for Authors

Memoirs are printed from camera copy fully prepared by the author. This means that the finished book will look exactly like the copy submitted.

The paper must contain a *descriptive title* and an *abstract* that summarizes the article in language suitable for workers in the general field (algebra, analysis, etc.). The *descriptive title* should be short, but informative; useless or vague phrases such as "some remarks about" or "concerning" should be avoided. The *abstract* should be at least one complete sentence, and at most 300 words. Included with the footnotes to the paper should be the 2000 *Mathematics Subject Classification* representing the primary and secondary subjects of the article. The classifications are accessible from `www.ams.org/msc/`. The list of classifications is also available in print starting with the 1999 annual index of *Mathematical Reviews*. The Mathematics Subject Classification footnote may be followed by a list of *key words and phrases* describing the subject matter of the article and taken from it. Journal abbreviations used in bibliographies are listed in the latest *Mathematical Reviews* annual index. The series abbreviations are also accessible from `www.ams.org/publications/`. To help in preparing and verifying references, the AMS offers MR Lookup, a Reference Tool for Linking, at `www.ams.org/mrlookup/`. When the manuscript is submitted, authors should supply the editor with electronic addresses if available. These will be printed after the postal address at the end of the article.

Electronically prepared manuscripts. The AMS encourages electronically prepared manuscripts, with a strong preference for $\mathcal{A}_\mathcal{M}\mathcal{S}$-LaTeX. To this end, the Society has prepared $\mathcal{A}_\mathcal{M}\mathcal{S}$-LaTeX author packages for each AMS publication. Author packages include instructions for preparing electronic manuscripts, the *AMS Author Handbook*, samples, and a style file that generates the particular design specifications of that publication series. Though $\mathcal{A}_\mathcal{M}\mathcal{S}$-LaTeX is the highly preferred format of TeX, author packages are also available in $\mathcal{A}_\mathcal{M}\mathcal{S}$-TeX.

Authors may retrieve an author package from e-MATH starting from `www.ams.org/tex/` or via FTP to `ftp.ams.org` (login as `anonymous`, enter username as password, and type `cd pub/author-info`). The *AMS Author Handbook* and the *Instruction Manual* are available in PDF format following the author packages link from `www.ams.org/tex/`. The author package can be obtained free of charge by sending email to `pub@ams.org` (Internet) or from the Publication Division, American Mathematical Society, 201 Charles St., Providence, RI 02904, USA. When requesting an author package, please specify AMS-LaTeX or AMS-TeX, Macintosh or IBM (3.5) format, and the publication in which your paper will appear. Please be sure to include your complete mailing address.

Sending electronic files. After acceptance, the source file(s) should be sent to the Providence office (this includes any TeX source file, any graphics files, and the DVI or PostScript file).

Before sending the source file, be sure you have proofread your paper carefully. The files you send must be the EXACT files used to generate the proof copy that was accepted for publication. For all publications, authors are required to send a printed copy of their paper, which exactly matches the copy approved for publication, along with any graphics that will appear in the paper.

TeX files may be submitted by email, FTP, or on diskette. The DVI file(s) and PostScript files should be submitted only by FTP or on diskette unless they are encoded properly to submit through email. (DVI files are binary and PostScript files tend to be very large.)

Electronically prepared manuscripts can be sent via email to `pub-submit@ams.org` (Internet). The subject line of the message should include the publication code to identify it as a Memoir. TeX source files, DVI files, and PostScript files can be transferred over the Internet by FTP to the Internet node `e-math.ams.org` (130.44.1.100).

Electronic graphics. Comprehensive instructions on preparing graphics are available at `www.ams.org/jourhtml/graphics.html`. A few of the major requirements are given here.

Submit files for graphics as EPS (Encapsulated PostScript) files. This includes graphics originated via a graphics application as well as scanned photographs or other computer-generated images. If this is not possible, TIFF files are acceptable as long as they can be opened in Adobe Photoshop or Illustrator. No matter what method was used to produce the graphic, it is necessary to provide a paper copy to the AMS.

Authors using graphics packages for the creation of electronic art should also avoid the use of any lines thinner than 0.5 points in width. Many graphics packages allow the user to specify a "hairline" for a very thin line. Hairlines often look acceptable when proofed on a typical laser printer. However, when produced on a high-resolution laser imagesetter, hairlines become nearly invisible and will be lost entirely in the final printing process.

Screens should be set to values between 15% and 85%. Screens which fall outside of this range are too light or too dark to print correctly. Variations of screens within a graphic should be no less than 10%.

Inquiries. Any inquiries concerning a paper that has been accepted for publication should be sent directly to the Electronic Prepress Department, American Mathematical Society, 201 Charles St., Providence, RI 02904, USA.

Editors

Titles in This Series

815 **Martin Bendersky and Donald M. Davis,** V_1-periodic homotopy groups of $SO(n)$, 2004

814 **Johannes Huebschmann,** Kähler spaces, nilpotent orbits, and singular reduction, 2004

813 **Jeff Groah and Blake Temple,** Shock-wave solutions of the Einstein equations with perfect fluid sources: Existence and consistency by a locally inertial Glimm scheme, 2004

812 **Richard D. Canary and Darryl McCullough,** Homotopy equivalences of 3-manifolds and deformation theory of Kleinian groups, 2004

811 **Ottmar Loos and Erhard Neher,** Locally finite root systems, 2004

810 **W. N. Everitt and L. Markus,** Infinite dimensional complex symplectic spaces, 2004

809 **J. T. Cox, D. A. Dawson, and A. Greven,** Mutually catalytic super branching random walks: Large finite systems and renormalization analysis, 2004

808 **Hagen Meltzer,** Exceptional vector bundles, tilting sheaves and tilting complexes for weighted projective lines, 2004

807 **Carlos A. Cabrelli, Christopher Heil, and Ursula M. Molter,** Self-similarity and multiwavelets in higher dimensions, 2004

806 **Spiros A. Argyros and Andreas Tolias,** Methods in the theory of hereditarily indecomposable Banach spaces, 2004

805 **Philip L. Bowers and Kenneth Stephenson,** Uniformizing dessins and Belyĭ maps via circle packing, 2004

804 **A. Yu Ol'shanskii and M. V. Sapir,** The conjugacy problem and Higman embeddings, 2004

803 **Michael Field and Matthew Nicol,** Ergodic theory of equivariant diffeomorphisms: Markov partitions and stable ergodicity, 2004

802 **Martin W. Liebeck and Gary M. Seitz,** The maximal subgroups of positive dimension in exceptional algebraic groups, 2004

801 **Fabio Ancona and Andrea Marson,** Well-posedness for general 2×2 systems of conservation law, 2004

800 **V. Poénaru and C. Tanas,** Equivariant, almost-arborescent representation of open simply-connected 3-manifolds; A finiteness result, 2004

799 **Barry Mazur and Karl Rubin,** Kolyvagin systems, 2004

798 **Benoît Mselati,** Classification and probabilistic representation of the positive solutions of a semilinear elliptic equation, 2004

797 **Ola Bratteli, Palle E. T. Jorgensen, and Vasyl' Ostrovs'kyĭ,** Representation theory and numerical AF-invariants, 2004

796 **Marc A. Rieffel,** Gromov-Hausdorff distance for quantum metric spaces/Matrix algebras converge to the sphere for quantum Gromov-Hausdorff distance, 2004

795 **Adam Nyman,** Points on quantum projectivizations, 2004

794 **Kevin K. Ferland and L. Gaunce Lewis, Jr.,** The $RO(G)$-graded equivariant ordinary homology of G-cell complexes with even-dimensional cells for $G = \mathbb{Z}/p$, 2004

793 **Jindřich Zapletal,** Descriptive set theory and definable forcing, 2004

792 **Inmaculada Baldomá and Ernest Fontich,** Exponentially small splitting of invariant manifolds of parabolic points, 2004

791 **Eva A. Gallardo-Gutiérrez and Alfonso Montes-Rodríguez,** The role of the spectrum in the cyclic behavior of composition operators, 2004

790 **Thierry Lévy,** Yang-Mills measure on compact surfaces, 2003

789 **Helge Glöckner,** Positive definite functions on infinite-dimensional convex cones, 2003

788 **Robert Denk, Matthias Hieber, and Jan Prüss,** $\mathcal{R}$-boundedness, Fourier multipliers and problems of elliptic and parabolic type, 2003

For a complete list of titles in this series, visit the
AMS Bookstore at **www.ams.org/bookstore/**.